Advances in Experimental Medicine and Biology

Neuroscience and Respiration

Volume 1114

Subseries Editor
Mieczyslaw Pokorski

More information about this subseries at http://www.springer.com/series/13457

Mieczyslaw Pokorski
Editor

Clinical Pulmonary Research

Springer

Editor
Mieczyslaw Pokorski
Opole Medical School
Opole, Poland

ISSN 0065-2598 ISSN 2214-8019 (electronic)
Advances in Experimental Medicine and Biology
ISBN 978-3-030-03869-4 ISBN 978-3-030-03870-0 (eBook)
https://doi.org/10.1007/978-3-030-03870-0

Library of Congress Control Number: 2018964104

This Springer imprint is published by the registered company Springer Nature Switzerland AG
The registered company address is: Gewerbestrasse 11, 6330 Cham, Switzerland

Preface

The book series Neuroscience and Respiration presents contributions by expert researchers and clinicians in the multidisciplinary areas of medical research and clinical practice. Particular attention is focused on pulmonary disorders as the respiratory tract is up front at the first line of defense for organisms against pathogens and environmental or other sources of toxic or disease-causing effects. The articles provide timely overviews of contentious issues or recent advances in the diagnosis, classification, and treatment of the entire range of diseases and disorders, both acute and chronic. The texts are thought as a merger of basic and clinical research dealing with biomedicine at both the molecular and functional levels and with the interactive relationship between respiration and other neurobiological systems, such as cardiovascular function, immunogenicity, endocrinology and humoral regulation, and the mind-to-body connection. The authors focus on modern diagnostic techniques and leading-edge therapeutic concepts, methodologies, and innovative treatments. The action and pharmacology of existing drugs and the development and evaluation of new agents are the heady area of research. Practical, data-driven options to manage patients are considered. New research is presented regarding older drugs, performed from a modern perspective or from a different pharmacotherapeutic angle. The introduction of new drugs and treatment approaches in both adults and children is also discussed.

Body functions, including lung ventilation and its regulation, are ultimately driven by the brain. However, neuropsychological aspects of disorders are still mostly a matter of conjecture. After decades of misunderstanding and neglect, emotions have been rediscovered as a powerful modifier or even the probable cause of various somatic disorders. Today, the link between stress and health is undeniable. Scientists accept a powerful psychological connection that can directly affect our quality of life and health span. Psychological approaches, which can decrease stress, can play a major role in disease therapy.

Neuromolecular and carcinogenetic aspects relating to gene polymorphism and epigenesis, involving both heritable changes in the nucleotide sequence and functionally relevant changes to the genome that do not involve a change in the nucleotide sequence, leading to disorders, are also tackled.

Clinical advances stemming from molecular and biochemical research are but possible if research findings are translated into diagnostic tools, therapeutic procedures, and education, effectively reaching physicians and patients. All this cannot be achieved without a multidisciplinary, collaborative, bench-to-bedside approach involving both researchers and clinicians. The role of science in shaping medical knowledge and transforming it into practical care is undeniable.

Concerning respiratory disorders, their societal and economic burden has been on the rise worldwide, leading to disabilities and shortening of life-span. COPD alone causes more than three million deaths globally each year. Concerted efforts are required to improve this situation, and part of those efforts are gaining insights into the underlying mechanisms of disease and staying abreast with the latest developments in diagnosis and treatment regimens. It is hoped that the articles published in this series will assume a leading position as a source of information on interdisciplinary medical research advancements, addressing the needs of medical professionals and allied health-care workers, and become a source of reference and inspiration for future research ideas.

I would like to express my deep gratitude to Paul Roos, and Cynthia Kroonen of Springer Nature NL for their genuine interest in making this scientific endeavor come through and in the expert management of the production of this novel book series.

Mieczyslaw Pokorski

Contents

Advs Exp. Medicine, Biology - Neuroscience and Respiration (2018) 41: 1–9
https://doi.org/10.1007/5584_2018_200

Published online: 12 May 2018

Coexistence of Chronic Bronchitis in Chronic Obstructive Lung Disease

Filip Mejza, Paweł Nastałek, Agnieszka Mastalerz-Migas, Zbigniew Doniec, and Wojciech Skucha

Abstract

The incidence of chronic obstructive pulmonary disease (COPD) is on the rise worldwide. Chronic bronchitis is a frequent accompaniment of COPD, which increases the burden of COPD in affected individuals. The aim of this study was to characterize the phenotype of chronic bronchitis in COPD patients. The study was based on the survey data retrospectively retrieved from the Action Health–Lung Cancer Prophylaxis and Health Care Improvement screening program that concerned all the inhabitants, aged over 40, of the Proszowice administrative region situated in the Lesser Poland Voivodeship in southern Poland. Participants with the symptoms suggestive of a lung disease were subject to further evaluation. The findings were that 546 (13.3%) out of the 4105 individuals displayed spirometry features of COPD. Symptoms of chronic bronchitis were present in 92 (16.8%) out of the COPD afflicted persons. Chronic bronchitis was commoner in current smokers and its incidence increased with increasing severity of airway obstruction. In multivariate analysis, chronic bronchitis was independently related to lower FEV1, FVC, FEV1/FVC, and to dyspnea. In regression model, factors related to increased risk of chronic bronchitis were current smoking, asthma, and lower lung function. We conclude that COPD with coexisting chronic bronchitis is linked to severer dyspnea and worse lung function. Current smoking, asthma, and lower lung function are related to increased risk of chronic bronchitis accompanying COPD.

Keywords

Airway obstruction · Asthma · Chronic bronchitis · COPD · Dyspnea · Lung disease · Screening

1 Introduction

Chronic obstructive pulmonary disease (COPD) is one of the most common chronic respiratory

F. Mejza (✉)
Center for Evidence Based Medicine, Second Department of Internal Medicine, Jagiellonian University Medical College, Cracow, Poland
e-mail: filipmejza@hotmail.com

P. Nastałek
Department of Pulmonology, Second Department of Internal Medicine, Jagiellonian University Medical College, Cracow, Poland

A. Mastalerz-Migas
Department of Family Medicine, Wrocław Medical University, Wrocław, Poland

Z. Doniec
Department of Pneumology, Institute of Tuberculosis and Lung Diseases, Rabka, Poland

W. Skucha
Pneumology Ward, Regional Hospital in Proszowice, Proszowice, Poland

conditions (Buist et al. 2007). It is associated with a significant burden in affected individuals (López-Campos et al. 2016). The majority of cases in developed countries are related to exposure to tobacco smoke, but the disease is heterogeneous, with emphysema and chronic bronchitis recognized as the most common phenotypes (GOLD 2017). Chronic bronchitis is usually defined as cough and sputum production for at least 3 months *per* year for at least two consecutive years (Definition and classification of chronic bronchitis 1965). Chronic bronchitis is common in COPD patients. However, available estimates of chronic bronchitis prevalence across COPD patients vary widely from 15 to more than 50% (De Oca et al. 2012; Agusti et al. 2010). A growing number of evidence suggest an increased disease burden in COPD patients with chronic bronchitis. These patients have worse quality of life, more exacerbations, and increased risk of death (Kim et al. 2011; Kanervisto et al. 2010; Ekberg-Aronsson et al. 2005). Recent papers also report an accelerated lung function decline in patient with chronic bronchitis, suggesting the bronchitis be a risk factor for COPD development (Guerra et al. 2009; Stavem et al. 2006; Vestbo et al. 1996). Despite these data, there is a paucity of studies focusing on the chronic bronchitis phenotype in COPD subjects.

The Proszowice region is primarily a rural region, located in southern Poland. The prevalence of COPD in the region is as high as 22.1% (10.9% of GOLD Spirometry Stage 2 and higher) (Nizankowska-Mogilnicka et al. 2007). There is also a relatively high prevalence of lung cancer reported in the county, compared to other parts of the region. For these reasons a large health-prophylaxis program, aimed to improve respiratory health status of region dwellers, has recently been introduced in the county (Skucha et al. 2017). The program includes several educational activities and a screening for COPD, comprising questionnaire, spirometry, and chest X-ray examinations. The aim of this study was to characterize the phenotype of chronic bronchitis in patients with spirometry confirmed COPD in this region of Poland.

2 Methods

The ethics approval to conduct the study was obtained from the Center for Evidence-Based Medicine, Second Department of Internal Medicine, Jagiellonian University Medical College in Cracow. The study was based on the data retrospectively retrieved from the Action Health–Lung Cancer Prophylaxis and Health Care Improvement screening program aimed at all inhabitants of the Proszowice administrative region situated in the Lesser Poland Voivodeship in southern Poland (Skucha et al. 2017). In brief, the program was based on a survey that included items on respiratory symptoms (including chronic cough and sputum production), previous diseases, tobacco or e-cigarettes smoking, and occupational hazard with emphasis on farming. All inhabitants of the region, aged over 40, were invited to participate. The survey was conducted in the face-to-face routine at the subject's homes by pre-trained staff and results were assessed by a medical coordinator of the program, according to the pre-established scoring rules. Patients with any symptoms suggestive of lung diseases, for instance, cough, sputum expectoration, dyspnea, or hemoptysis, or those with significant exposure to inhaled noxious agents were examined using spirometry and chest X-rays, subjected to further diagnostic procedure if required. Spirometry was performed with the post-bronchodilator test using 400 μg of inhaled salbutamol (LungTest 1000 spirometer; MES; Cracow, Poland). COPD was diagnosed based on the post-bronchodilator FEV1/FVC ratio below a lower limit of the norm for age and gender. The modified Medical Research Council (mMRC) dyspnea scale was used to assess the functional impairment attributable to dyspnea in COPD patients (Mahler and Wells 1988). Chronic cough or chronic sputum expectoration was defined as an affirmative answer to the question: 'Do you cough or expel phlegm on most days for at least 3 months each year', respectively. Chronic bronchitis was diagnosed in COPD patients who reported both chronic cough and phlegm for at least 2 consecutive years. Asthma and other medical diagnoses were assessed based on a self-reported survey.

2.1 Statistical Evaluation

Data are presented as means ±SD, medians with interquartile ranges (IQR), or proportions as indicated. Student's *t*-test, Mann-Whitney U test, and Chi-square or Fisher's exact test were used for inter-group comparisons. To verify the hypothesis that chronic bronchitis would be related to worse lung function and greater dyspnea, we constructed the following two regression models: 1/ with the post-bronchodilator FEV1, FVC, and FEV1/FVC as dependent variables and chronic bronchitis, gender, age, smoking status, and self-reported asthma as potential predictors; and 2/ with the mMRC as a dependent variable and chronic bronchitis, gender, age, smoking status, self-reported asthma and FEV1 as potential predictors. Potential risk factors for chronic bronchitis were examined by multiple logistic regression models, aside from chronic bronchitis including the following covariates: basic demographic data (age, gender, and education), smoking status, e-cigarettes use, exposure to passive smoking, occupational exposures (working in mining and metallurgy industry, welding, farming, and using pesticides or fertilizers), home heating systems (coal, coke gas, or wood), way of cooking at home (gas or electric oven; wood or coal fuelled stove), self-reported asthma, evidence of the presence of bronchiectasis, and the FEV1 value in post-bronchodilator test. A p-value <0.05 defined statistically significant differences. The analysis was performed using R statistical v3.3.1 software (https://www.R-project.org).

3 Results

Overall, data were available from 14,323 subjects (98.4% returns of questionnaires and 70.1% of target population of all inhabitants of the region aged ≥40). Four thousand one hundred and five subjects provided pre- and post-bronchodilator spirometry of high quality (28.7% of those with questionnaire data available and 20.3% of the inhabitants). Of those, 547 (13.3%) subjects had irreversible airway obstruction in spirometry (post-bronchodilator FEV1/FVC below the lower limit of normal), pointing to the presence of COPD. These subjects were included in further analysis. Chronic bronchitis was present in 92 (16.8%) out of the COPD patients; this proportion increased with the severity of airway obstruction. Component symptoms of chronic bronchitis – chronic cough and chronic sputum expectoration – were more commonly manifest in these patients. A proportions of subjects reporting these symptoms also increased with the severity of airway obstruction (Fig. 1). Basic characteristics of the patients studied are presented in Table 1. Symptoms of chronic bronchitis lasted for 2–5 years in 34.8% of the patients and for more than 5 years in the remaining 65.2% of the patients. Symptoms were commoner in men and in current smokers.

COPD patients with coexisting chronic bronchitis had a greater dyspnea score and worse lung function (Table 2). Also, patients who presented with chronic cough or sputum expectoration tended to have worse lung function and stronger dyspnea, compared with the patients not manifesting these symptoms (Fig. 2).

Patients with chronic bronchitis were more often diagnosed with COPD or asthma, and with non-respiratory comorbidities. In the multivariate analysis, after adjustment for gender, age, smoking status, and self-reported asthma, chronic bronchitis remained independently related to lower FEV1 (−6.7%, 95%CI: -11.2–(−2.1); $p = 0.004$), lower FVC (−6.3%, 95%CI: -11.9–(−0.8); $p = 0.026$), and lower FEV1/FVC (−1,8, 95%CI: -3.4–(−0.2); $p < 0.029$). Likewise, relationship of chronic bronchitis to the severity of dyspnea was independent from other confounders (mMRC 0.8, 95%CI: 0.5–1.2; $p < 0.001$). In the regression model, the following factors were related to an increased risk of reporting chronic bronchitis: current smoking, self-reported asthma, and lower lung function expressed as a 10% reduction in FEV1 (Table 3).

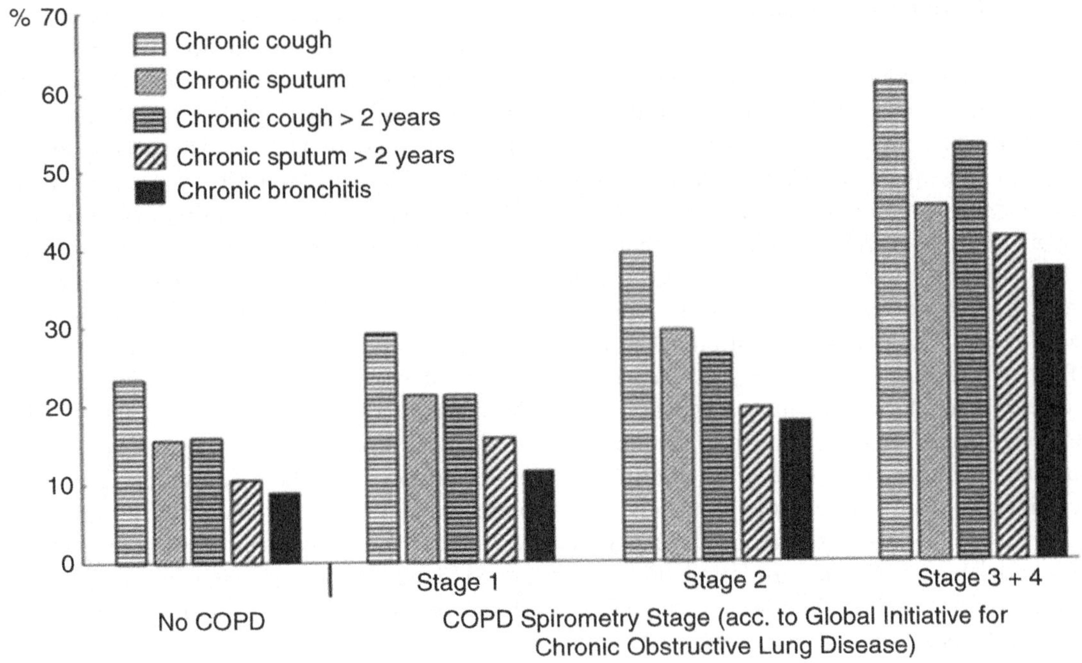

Fig. 1 Prevalence of chronic cough, chronic sputum expectoration, chronic bronchitis in patients without and with chronic obstructive pulmonary disease (COPD). COPD severity was based on the spirometry stage, according to the Global Initiative for Chronic Obstructive Lung Diseases (GOLD 2017)

Table 1 Basic patient characteristics

	Total (*n* = *547*)	CB+ (*n* = *92*)	CB- (*n* = *455*)	*p*-value CB+ *vs.* CB-
Age; year (mean ± SD)	63.9 ± 10.9	64.7 ± 9.9	63.8 ± 11.1	0.442
Women; n (%)	273 (49.9)	31 (33.7)	242 (53.2)	0.007
Education; n (%)				
≤8 years (primary school)	339 (65.2)	63 (70.8)	276 (64.0)	0.134
9–12 years (high school)	163 (31.4)	21 (23.6)	142 (33.0)	
>12 years (university)	18 (3.5)	5 (5.6)	13 (3.0)	
Smoking; n (%)				
Never-smokers	177 (32.7)	15 (16.5)	162 (36.0)	<0.001
Ex-smokers	171 (31.6)	27 (29.7)	144 (32.0)	
Current smokers	193 (35.7)	49 (53.9)	144 (32.0)	
Current passive exposure to tobacco smoke; n (%)	126 (23.2)	23 (25.0)	103 (22.8)	0.654

CB+, chronic bronchitis present; CB-, chronic bronchitis absent. The number of patients in the *Education* and *Smoking* categories does not sum up to that given in the headline due to lack of data in some cases

4 Discussion

Our findings demonstrate that COPD combined with chronic bronchitis presents a greater disease burden, with worse lung function and severer dyspnea, compared to COPD alone. Patients with coexisting chronic bronchitis are more often current smokers. Yet a direct relationship of chronic bronchitis with worse lung function is disputable due to a cross-sectional nature of this study. The seminal studies of Fletcher and Peto (1977) and Peto et al. (1983) have demonstrated

Table 2 Dyspnea, lung function, and medical history

	Total (*n* = *547*)	CB+ (*n* = *92*)	CB- (*n* = *455*)	*p*-value CB+ *vs.* CB-
Dyspnea; n (%)	265 (48.5)	70 (76.1)	195 (42.9)	<0.001
mMRC; mean ± SD	0.5 ± 0.1	1.7 ± 2.0	0.3 ± 0.1	<0.001
Lung function[a]				
FEV1; % predicted; mean ± SD	77.8 ± 21.2	68.8 ± 20.2	79.6 ± 20.9	<0.001
FVC; % predicted; median (IQR)	103.5 (88.4–119.9)	94.6 (81.3–110.1)	104.7 (89.5–121.2)	<0.001
FEV_1/%FVC; median (IQR)	61.0 (55.0–64.7)	58.3 (50.7–62.5)	61.3 (55.7–64.9)	0.001
Diagnosis				
COPD; n (%)	53 (9.7)	24 (26.1)	29 (6.4)	p < 0.001
Emphysema; n (%)	25 (4.6)	8 (8.7)	17 (3.7)	p = 0.052
COPD and/or emphysema; n (%)	65 (11.9)	26 (28.3)	39 (8.6)	p < 0.001
Asthma; n (%)	117 (21.4)	35 (38.0)	82 (18.0)	p < 0.001
Comorbidities[b]; n (%)	416 (76.1)	78 (84.8)	338 (74.3)	0.031
Hospitalizations for respiratory problems year past; mean ± SD	0.1 ± 0.5	0.2 ± 0.9	0.1 ± 0.3	0.016

CB+, chronic bronchitis present; CB-, chronic bronchitis absent
[a]post-bronchodilator; [b]self-reported chronic non-pulmonary diseases

that chronic bronchitis does not influence the risk of airway obstruction or death in chronic obstructive airway disease. However, ample later research has brought divergent, suggesting the presence of chronic cough and sputum be related to negative outcomes. Prospective observations have shown that chronic bronchitis is related to accelerated rate of lung function decline and COPD development (Guerra et al. 2009; Ekberg-Aronsson et al. 2005; Vestbo et al. 1996). Yet more extended longitudinal data are required to make a decisive judgment of whether chronic bronchitis is a risk factor for more severe course and progression of COPD or is rather a consequence of more severe airway obstruction and worsened lung function. Nonetheless, the presence of chronic bronchitis is also related to other negative outcomes, independently from lung function or smoking status, such as exaggerated dyspnea, greater number of comorbidities, and more frequent hospitalizations for respiratory problems. Negative outcomes reported in COPD subjects with chronic bronchitis include worse quality of life and increased risk of respiratory exacerbations (Mejza et al. 2017; Kanervisto et al. 2010). Part of the exacerbation burden may be attributed to co-existent asthma. In the present study, asthma was reported in more than one third of COPD patients with chronic bronchitis. Asthma-COPD overlap is evidently related to worse health status (Postma and Rabe 2015). The relationships are independent in a multivariate analysis that includes asthma as a cofactor, and are still present when patients reporting asthma are removed from the sample studied. However, both the present data and those reported in other studies on the co-existence of asthma may be biased as asthma is usually a self-reported diagnosis in epidemiological reports. A proportion of patients with asthma symptoms among COPD patients is probably as high as 15–45% (Postma and Rabe 2015). Thus, there is a need for improved classification of chronic airway diseases.

Published estimates of the proportion of chronic bronchitis in COPD patients vary from 7% to over 50% (De Oca et al. 2012; Kim et al. 2011; Agusti et al. 2010; Halbert et al. 2006). Such a wide range reflects not only differences in the prevalence of chronic bronchitis, but also different diagnosing and reporting methods used in various studies. The variability foremost

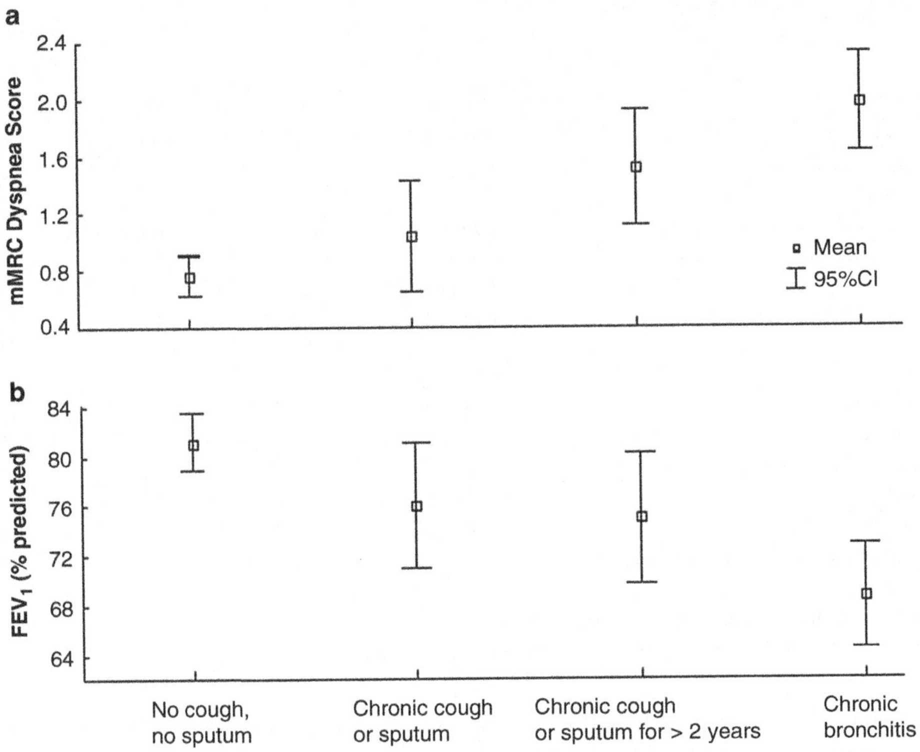

Fig. 2 Dyspnea (**a**) and lung function (**b**) in chronic obstructive pulmonary disease (COPD) patients manifesting different symptoms

concerns the way in which patient populations were recruited as it is known that symptoms of chronic bronchitis are more common in those patients who run a severe COPD course or in selected populations of treated patients. The estimated proportion of chronic bronchitis coexisting with COPD found in the present study is placed near the lower end of the range above outlined and fits well into data from large international studies such as BOLD, estimated proportion of 10% based on data from 29 countries, and PLATINO, estimated proportion of 14.4% based on the data from five South-American countries (Mejza et al. 2017; De Oca et al. 2012).

The most commonly used definition of chronic bronchitis, i.e., cough and sputum production for at least 3 months for two or more consecutive years, is arbitrary. Expectedly, more COPD patients reported chronic cough or chronic sputum than chronic bronchitis. Patients who reported the symptoms of chronic cough or sputum expectoration, but failed the exact criteria for chronic bronchitis, still had worse lung function and more severe dyspnea compared to those without these symptoms. In line with the results of a study by Kim et al. (2016), we found that two thirds of patients with chronic bronchitis had symptoms lasting for a time longer than 5 years. Of note, most patients with COPD alone do not report chronic cough or sputum, even among those with severe airway obstruction. We noticed that the antecedent diagnosis of COPD had been made just in 12% of patients, and the diagnosis was probably facilitated by the presence of symptoms of chronic bronchitis as a proportion

Table 3 Risk factors for chronic bronchitis in multivariate analysis

Potential risk factors		OR	95% CI	*p*-value
Male sex		1.47	0.78–2.78	0.23
Age; years	30–49	1	Ref.	–
	50–59	0.77	0.26–2.24	0.626
	60–69	1.26	0.48–3.31	0.645
	70+	0.98	0.34–2.82	0.967
Education	Other than university	1	Ref.	–
	University	3.14	0.85–11.58	0.086
Smoking status	Never smokers	1	Ref.	–
	Former smokers	1.65	0.76–3.60	0.210
	Current smokers	3.48	1.59–7.60	0.002
E-cigarettes use		2.01	0.81–5.00	0.133
Exposure to passive smoke		1.06	0.58–1.93	0.852
Working for >3 months in	Coal mine	3.64	0.75–17.63	0.109
	Stealworks	0.42	0.11–1.67	0.221
	Welding industry	0.82	0.31–2.18	0.691
	Farming	1.43	0.61–3.36	0.418
Occupational exposure to pesticides or fertilizers		0.89	0.44–1.80	0.741
Occupational exposure to asbestos		1.66	0.26–10.59	0.590
Home heating with	Coal	1.21	0.31–4.74	0.790
	Coke	1.87	0.17–20.56	0.611
	Gas	2.68	0.61–11.74	0.190
	Wood	1.12	0.62–2.00	0.712
Cooking with	Gas	1	Ref.	–
	Electric	0.99	0.24–4.16	0.993
	Wood	0.55	0.18–1.73	0.304
	Coal	1.09	0.08–14.21	0.946
Asthma		3.19	1.77–5.74	<0.001
Bronchiectasis		0.55	0.06–5.48	0.610
10% FEV1 reduction		1.21	1.05–1.39	0.006

OR, odds ratio; 95% CI, 95% confidence intervals

of properly diagnosed COPD was smaller in patients without such symptoms. These observations speak in favor of spirometry screening, although it is currently not viewed as an effective tool to predict the health status of a population (US Preventive Services Task Force et al. 2016). In the present study, a significant proportion of patients with previously undiagnosed COPD failed to manifest the symptoms of chronic cough and sputum. Further, many patients appeared to have severe COPD, in which pharmacologic interventions could potentially have been of benefit.

Patients with chronic bronchitis were more often diagnosed with COPD or asthma. They also more often reported non-respiratory comorbidities. That observation may stem from a more frequent contact with healthcare providers or by a relation of chronic bronchitis to worse lung function and hence worse general health status. The risk factors for chronic bronchitis identified in this study are in line with those reported in other studies that have demonstrated asthma and current smoking to be notably important (Mejza et al. 2017; Doney et al. 2014). Neither occupational exposures nor in-house potential risk factors were appreciably related to increased risk of chronic bronchitis.

This study has some drawbacks. Firstly, the sample of patients for spirometry was not randomly chosen, which might have overestimated the prevalence of chronic bronchitis among COPD sufferers. However, the size of the study population may be an offsetting factor concerning this limitation as we invited all inhabitants, aged over 40, of a defined administrative region to participate in the study, getting a response rate as high as 70%, and spirometry was performed in more than 20% of region inhabitants. Secondly, a cross-sectional nature of the study makes the causal analysis of the identified relationships impossible. Nonetheless, we believe that the strength of the study, apart from the population sample, also lies in a strict quality control of data acquisition.

We conclude that symptoms of chronic bronchitis are reported by a minority of COPD patients and the presence of such symptoms is related to coexistence of asthma and current smoking. Chronic bronchitis identifies patients having a more severe COPD.

Acknowledgments Funded by Norway Grants – Program PL 13 entitled 'Reducing social inequalities in health'. The founder played no role in planning and executing this study.

Conflicts of Interest The authors declare no conflicts of interest in relation to this article.

References

Agusti A, Calverley PM, Celli B, Coxson HO, Edwards LD, Lomas DA, MacNee W, Miller BE, Rennard S, Silverman EK, Tal-Singer R, Wouters E, Yates JC, Vestbo J, Evaluation of COPD Longitudinally to Identify Predictive Surrogate Endpoints (ECLIPSE) Investigators (2010) Characterisation of COPD heterogeneity in the ECLIPSE cohort. Respir Res 11:122

Buist AS, MA MB, Vollmer WM, Gillespie S, Burney P, Mannino DM, Menezes AM, Sullivan SD, Lee TA, Weiss KB, Jensen RL, Marks GB, Gulsvik A, Nizankowska-Mogilnicka E, BOLD Collaborative Research Group (2007) International variation in the prevalence of COPD (the BOLD study): a population-based prevalence study. Lancet 370:741–750

De Oca MM, Halbert RJ, Lopez MV, Perez-Padilla R, Tálamo C, Moreno D, Muiño A, Jardim JR, Valdivia G, Pertuzé J, Menezes AM (2012) The chronic bronchitis phenotype in subjects with and without COPD: the PLATINO study. Eur Respir J 40:28–36

Definition and classification of chronic bronchitis for clinical and epidemiological purposes. A report to the Medical Research Council by their Committee on the Aetiology of Chronic Bronchitis (1965) Lancet 1:775–779

Doney B, Hnizdo E, Graziani M, Kullman G, Burchfiel C, Baron S, Fujishiro K, Enright P, Hankinson JL, Stukovsky KH, Martin CJ, Donohue KM, Barr RG (2014) Occupational risk factors for COPD phenotypes in the multi-ethnic study of atherosclerosis (MESA) lung study. COPD 11:368–380

Ekberg-Aronsson M, Pehrsson K, Nilsson JA, Nilsson PM, Löfdahl CG (2005) Mortality in GOLD stages of COPD and its dependence on symptoms of chronic bronchitis. Respir Res 6:98

Fletcher C, Peto R (1977) The natural history of chronic airflow obstruction. BMJ 1:1645–1648

GOLD (2017) From the global strategy for the diagnosis, management and prevention of COPD, Global initiative for chronic obstructive lung disease. http://goldcopd.org. Accessed on 1 Feb 2017

Guerra S, Sherrill DL, Venker C, Ceccato CM, Halonen M, Martinez FD (2009) Chronic bronchitis before age 50 years predicts incident airflow limitation and mortality risk. Thorax 64:894–900

Halbert RJ, Natoli JL, Gano A, Badamgarav E, Buist AS, Mannino DM (2006) Global burden of COPD: systematic review and meta-analysis. Eur Respir J 28:523–532

Kanervisto M, Saarelainen S, Vasankari T, Jousilahti P, Heistaro S, Heliövaara M, Luukkaala T, Paavilainen E (2010) COPD, chronic bronchitis and capacity for day-to-day activities: negative impact of illness on the health-related quality of life. Chron Respir Dis 7:207–215

Kim V, Han MK, ance GB, Make BJ, Newell JD, Hokanson JE, Hersh CP, Stinson D, Silverman EK, Criner GJ, Investigators COPDG (2011) The chronic bronchitic phenotype of COPD: an analysis of the COPDGene study. Chest 140:626–633

Kim V, Zhao H, Boriek AM et al (2016) Persistent and newly developed chronic bronchitis are associated with worse outcomes in chronic obstructive pulmonary disease. Ann Am Thorac Soc 13:1016–1025

López-Campos JL, Tan W, Soriano JB (2016) Global burden of COPD. Respirology 21:14–23

Mahler DA, Wells CK (1988) Evaluation of clinical methods for rating dyspnea. Chest 93:580–586

Mejza F, Gnatiuc L, Buist AS, Vollmer WM, Lamprecht B, Obaseki DO, Nastalek P, Nizankowska-Mogilnicka E, Burney PGJ, collaborators BOLD, study collaborators BOLD (2017) Prevalence and burden of chronic bronchitis symptoms: results from the BOLD study. Eur Respir J 50(5):1700621

Nizankowska-Mogilnicka E, Mejza F, Buist AS, Vollmer WM, Skucha W, Harat R, Pajak A, Gasowski J, Frey J, Nastalek P, Twardowska M, Janicka J, Szczeklik A (2007) Prevalence of COPD and tobacco smoking in Malopolska region–results from the BOLD study in Poland. Pol Arch Med Wewn 117:402–410

Peto R, Speizer FE, Cochrane AL, Moore F, Fletcher CM, Tinker CM, Higgins IT, Gray RG, Richards SM, Gilliland J, Norman-Smith B (1983) The relevance in adults of air-flow obstruction, but not of mucus hypersecretion, to mortality from chronic lung disease. Results from 20 years of prospective observation. Am Rev Respir Dis 128:491–500

Postma DS, Rabe KF (2015) The asthma-copd overlap syndrome. N Engl J Med 373:1241–1249

Skucha W, Mejza F, Nastalek P, Doniec Z (2017) Pulmonary prevention program in the Proszowice county: description and results. Adv Respir Med 85:239–245

Stavem K, Sandvik L, Erikssen J (2006) Can global initiative for chronic obstructive lung disease stage 0 provide prognostic information on long-term mortality in men? Chest 130:318–325

US Preventive Services Task Force (USPSTF), Siu AL, Bibbins-Domingo K, Grossman DC, Davidson KW, Epling JW Jr, García FA, Gillman M, Kemper AR, Krist AH, Kurth AE, Landefeld CS, Mangione CM, Harper DM, Phillips WR, Phipps MG, Pignone MP (2016) Screening for chronic obstructive pulmonary disease: US preventive services task force recommendation statement. JAMA 315:1372–1377

Vestbo J, Prescott E, Lange P (1996) Association of chronic mucus hypersecretion with FEV1 decline and chronic obstructive pulmonary disease morbidity. Copenhagen City heart study group. Am J Respir Crit Care Med 153:1530–1535

Advs Exp. Medicine, Biology - Neuroscience and Respiration (2018) 41: 11–17
https://doi.org/10.1007/5584_2018_206

Published online: 21 April 2018

Influence of Gaseous Pollutants on COPD Exacerbations in Patients with Cardiovascular Comorbidities

Michal Zielinski, Mariusz Gasior, Dariusz Jastrzebski, Aneta Desperak, and Dariusz Ziora

Abstract

Exacerbations of chronic obstructive pulmonary disease (COPD) are a serious public health issue. Ambient pollution and meteorological factors are considered among precipitating factors. There are few data concerning the impact of ambient pollutants other than particulates on COPD exacerbations. Among gaseous pollutants four main groups of substances are primarily monitored: nitrogen oxides (NOx), sulphur dioxide (SO_2), carbon monoxide (CO), and ozone (O_3). In this study, 12,889 hospitalizations in the years 2006–2014 due to exacerbations of COPD in patients having a co-existing cardiovascular pathology were retrospectively analyzed. Cardiovascular disease was ruled out as the underlying reason of hospitalization. Data concerning the then accompanying gaseous pollutants and weather conditions were collected. The findings were that the impact of SO_2 content was significantly associated with the relative risk (RR) of COPD exacerbation when the exposure took place at least 30 days or longer before hospital admission (RR 1.04–1.05; $p < 0.05$). In contrast, risk of COPD exacerbation rose when a shortening of the time lag between exposure to NOx and hospital admission was considered (RR 1.02–1.04; $p < 0.05$). O_3 exposure was associated with a lower risk irrespective of the length of exposure/exacerbation lag (RR 0.77–0.90; $p < 0.05$). There were insignificant associations observed for CO. In conclusion, the study demonstrates a salient influence of a co-existing cardiovascular malady on the appearance of COPD-related respiratory exacerbations when the pollutant SO_2 and NOx contents rose. In contrast, higher O_3 content was associated with a lower risk of COPD exacerbation.

Keywords

Cardiovascular comorbidities · COPD · Emergency admissions · Exacerbations · Gaseous air pollutants · Environmental health.

1 Introduction

COPD was the fifth leading cause of death in 2002 and it is predicted to become the third

M. Zielinski (✉), D. Jastrzebski, and D. Ziora
Department of Lung Diseases and Tuberculosis, School of Medicine with the Division of Dentistry in Zabrze, Medical University of Silesia, Katowice, Poland
e-mail: michal.zielinski1@interia.pl

M. Gasior and A. Desperak
Third Department of Cardiology, School of Medicine with the Division of Dentistry in Zabrze, Medical University of Silesia, Katowice, Poland

leading cause of death in the world by 2020 (Ko et al. 2016). Exacerbations of COPD are a serious public health issue as they are associated with significant mortality and morbidity. Exacerbations also play an important role in the course of the disease. They are notably characterized by a worse feeling of dyspnea, beyond normal day-to-day variability and may negatively influence patients' quality of life and accelerate pulmonary function decline. Moreover, COPD exacerbations account for most of hospitalizations and the associated economic burden (Donaldson et al. 2002).

Dyspnea, which is the defining symptom of a COPD exacerbation, is also frequently encountered in cardiovascular diseases. Congestive heart failure also may lead to obstructive or restrictive patterns in pulmonary function tests, emulating the findings of pulmonary pathologies (Hawkins et al. 2009). A similar clinical presentation can result in misdiagnosing COPD exacerbations as a cardiovascular problem. In addition, common risk factors which cause greater prevalence of COPD in patients with cardiovascular diseases make the differential diagnosis even more challenging (Kitzman et al. 2001). Co-existence of COPD with heart failure leads to a significant increase in mortality (Rushton et al. 2015).

COPD exacerbations are mainly caused by viral or bacterial lower respiratory tract infections, which constitute up to 81% of cases. Ambient pollution and meteorological factors are also considered among precipitating agents or cofactors (Woodhead et al. 2011). Although there are studies analyzing a connection between the particulate matter air pollution and the course of COPD, the data on the impact on COPD exacerbations of gaseous pollutants are scarce. According to our knowledge, there are no studies on the impact on COPD exacerbations of air pollutants in selected populations of patients with cardiovascular diseases. Gaseous pollutants are a heterogeneous group of air pollutants, classified according to their chemical structure. Due to their prevalence and potential threat to public health the following pollutants are notably monitored: sulphur dioxide (SO_2), carbon monoxide (CO), nitrogen oxides (NOx), and ozone (O_3) (Shah et al. 2013). The current study seeks to define the relationship between gaseous air pollutants and COPD exacerbations in patients with cardiovascular diseases, taking also into account meteorological data.

2 Methods

The study was approved by Ethics Committee of Medical University of Silesia; permit no. KNW/0022/KB/117/16. This is a retrospective study that concerns the health situation in the region of the Upper Silesian urban area of a population of 2 M people. The region accounts for 4% of the territory of Poland, but its industry produces 50% of gaseous pollutants emitted in the country. Data on weather conditions (air temperature, humidity, wind velocity, and atmospheric pressure) and gaseous air pollutants' concentrations (SO_2, CO, O_3, NOx, and separately NO and NO_2) were collected from the Upper Silesia air quality monitoring system database available on website of the Silesian regional inspectorate of environment. There were six air monitoring stations in the studied region located in the cities of Dąbrowa Górnicza, Gliwice, Katowice, Sosnowiec, Tychy, and Zabrze. The collected data encompassed the years 2006–2014.

Data concerning hospitalizations related to COPD exacerbations were obtained from the Silesian Cardiovascular Database, administered by the Science Department of the Silesian Center for Heart Disease in Zabrze and by the Department of Biostatistics of the Medical University of Silesia. The database covers hospitalizations of patients, with a diagnosis of a cardiovascular disease, treated in the wards of cardiology, cardiac surgery, vascular surgery, diabetology, internal medicine, and in the intensive care units in 310 Upper Silesian hospitals. This database is produced on the basis of reports collected from the Polish National Health Fund, the sole public health service provider in Poland. COPD exacerbation was defined as a hospitalization with the ICD-10 code J44 as the main diagnosis. Cardiovascular disease was ruled out as the underlying reason of hospitalization. There were 12,889 such hospitalizations in the period of time above outlined. The mean age of patients was 71 ± 10 years. Men accounted for 61.8% of the hospitalized patients. The database did not include outpatient visits, and therefore could create some bias due to exclusion of mild-to-moderate COPD exacerbations not necessitating hospitalization. Data concerning the presence of up to three co-morbid conditions were also evaluated (Table 1). The measurements of SO_2, CO, NOx, NO and NO_2, and O_3 were available for the whole period under consideration. The annual average concentration of these pollutants is presented in Table 2.

Statistical analysis was performed using the generalized linear model with a logarithmic link function. The following variables were assumed to be confounding factors: season, air temperature and humidity, wind velocity, and atmospheric pressure. Whole-day measurements for the Upper Silesian urban region were averaged. The period of up to 90 days before the day of hospital admission were considered. Statistical significance was assumed for $p < 0.05$.

3 Results

The relationship between the concentration of gaseous air pollutants and the risk of COPD exacerbation in the studied population of patients at the time of hospitalization and at different times preceding hospitalization is shown in Table 3. Relative risks (RR) were calculated *per* 1 interquartile range of pollutant concentration increase (IQR-P). Statistically significant associations were observed for all of the pollutants except CO. The effect of SO_2 concentration on COPD exacerbations became significant when a period of at least 30 days before hospitalization day was taken into account. The estimated effect increased with a longer timeframe, up to the longest period analyzed, suggesting a cumulative effect.

A statistically significant association was observed for all nitric oxides beginning from the day of hospital admission. The association was the greatest on the day of admission. In contrast to sulphur oxide, the effect was weakening and became statistically insignificant when longer time-frames were considered. The increase in ozone concentration was associated with a lower risk of COPD exacerbation. This protective effect was not observed for any other gaseous pollutant analyzed.

4 Discussion

This study evaluated the effects of gaseous pollutants on COPD exacerbations in patients with co-existing cardiovascular diseases. There are consistent data connecting the short- and long-term exposure to SO_2 and clinical respiratory worsening, particularly in patients with asthma. In contrast to asthma, available data on COPD exacerbation and SO_2 are limited. The present findings are in line with those of a study conducted in the London population, in which a 24-h SO_2 concentration was akin to the present one. The authors have reported the odds ratio 1.006 for 1 interquartile range change in SO_2. However, only 1-day time lags between exposure to SO_2 and the exacerbation were analyzed (Peacock et al. 2011). Likewise, a study from Canada analyzed the visits to emergency departments due to respiratory and cardiac conditions. There were no appreciable associations between exposure to SO_2 and COPD exacerbations and other respiratory diagnoses, despite relatively high content of SO_2 in some parts of the studied area. Again, however, lags no longer than 2 days were considered (Stieb et al. 2009). The present findings that a cumulative change in SO_2 content increases the extent of the association with COPD exacerbations and the emergency hospital admissions in the patients with a co-existing cardiovascular malady are consistent with a Brazilian study, conducted in the general population of COPD patients, in which longer exposure/exacerbation lags have been investigated (Arbex et al. 2009). Thus, studies which consider only the short-term effects can underestimate the actual frequency COPD exacerbations due to exposure to SO_2.

Table 1 Ten most frequent cardiovascular co-morbidities among the 12,889 hospitalizations

ICD-10	Comorbidity	Patients; n (%)
I10	Essential (primary) hypertension	3461 (26.9)
I50.9	Heart failure, unspecified	1370 (10.6)
I25.9	Chronic ischaemic heart disease, unspecified	1334 (10.4)
I25.0	Atherosclerotic cardiovascular disease	1038 (8.1)
I48	Atrial fibrillation and flutter	906 (7.0)
I11.9	Hypertensive heart disease without congestive heart failure	901 (7.0)
I70.9	Generalized and unspecified atherosclerosis	863 (6.7)
I50.0	Congestive heart failure	644 (5.0)
I25.8	Other forms of chronic ischemic heart disease	470 (3.7)
I25.5	Ischemic cardiomyopathy	397 (3.1)

Table 2 Concentration ($\mu g/m^3$) of gaseous pollutants during the period investigated

Year	SO_2	NO	NO_2	NOx	CO	O_3
2006	26.4 ± 28.3	17.1 ± 26.1	32.4 ± 16.9	58.5 ± 54.3	1.1 ± 0.9	41.4 ± 22.5
2007	14.4 ± 10.2	12.3 ± 17.0	26.3 ± 10.4	44.9 ± 34.2	0.8 ± 0.5	35.2 ± 17.5
2008	14.7 ± 10.1	11.8 ± 14.7	25.9 ± 10.3	43.7 ± 30.9	0.8 ± 0.4	37.3 ± 19.5
2009	16.1 ± 14.6	12.7 ± 16.1	26.6 ± 11.2	45.8 ± 33.6	0.9 ± 0.7	42.1 ± 23.0
2010	19.6 ± 18.6	14.6 ± 18.1	29.7 ± 13.9	52.1 ± 39.0	0.9 ± 0.7	38.4 ± 20.2
2011	17.1 ± 11.4	16.0 ± 21.2	33.7 ± 11.4	58.1 ± 41.7	1.1 ± 0.8	42.2 ± 19.9
2012	17.9 ± 17.4	13.7 ± 16.4	34.8 ± 12.5	55.7 ± 35.2	1.1 ± 0.8	42.2 ± 22.2
2013	14.4 ± 10.8	11.0 ± 10.1	27.8 ± 9.2	44.7 ± 22.4	0.8 ± 0.4	42.0 ± 21.3
2014	14.9 ± 9.5	10.8 ± 12.6	24.4 ± 9.6	40.8 ± 27.2	0.9 ± 0.5	38.0 ± 19.0

Data are means ±SD

In the ULTRA study (de Hartog et al. 2003), effects of gaseous pollutants in a group of elderly patients with coronary heart disease have been investigated. The results demonstrate just a trend toward a positive association between the CO concentration and the avoidance of activities. In contradistinction to the elderly patients, CO pollution has been associated with the presence of airway wheezes and congested nasal passages in children; however, CO concentration was averaged for 8 h, unlike 24 h in the present study (Rodriguez et al. 2007). Another two studies have investigated the association between CO and lung function reduction in the European population. In a Finish study, which included asthmatics, CO content amounted to half of that noticed in the region of the present study. This lower CO content has an inverse association with the morning and evening peak expiratory flow rates (Penttinen et al. 2001). On the other hand, in an Italian study, CO content in region of Rome surpassed that of the present study. The population investigated consisted of patients having single asthma, COPD, and ischemic heart disease. None of the patients had any co-existing respiratory or circulatory disease. There was no noticeable association between CO exposure, with a lag of up to 72 h, and forced vital capacity (FVC) or forced expired volume in 1 s (FEV1) in the groups investigated (Lagorio et al. 2006). The present findings in a subpopulation of COPD patients with cardiovascular co-morbidities confirm those made in the general population of COPD patients, and point to the lack of an association between CO content and COPD exacerbations.

Tropospheric O_3 is a pollutant with oxidative properties, formed in the photochemical reactions from precursor pollutants. Akin to SO_2, exposure to O_3 is associated with a worsening of the respiratory symptoms in asthma patients. It has been demonstrated in a study from New Jersey that although O_3 is connected with a higher frequency of asthma-related visits to the hospital, this

Table 3 Relative risk (RR) of COPD exacerbation depending on changes in pollutant concentration in different time-frames before hospitalization

	SO_2		NO		NO_2		NOx		CO		O_3	
Hospitalization time	p	RR	p	RR	p	RR	p	RR	p	RR	p	RR
At admission	0.56	1.01	<0.01	1.04	<0.01	1.14	<0.01	1.07	0.10	1.02	<0.01	0.77
3 days before admission	0.48	1.01	0.046	1.02	0.10	1.03	0.045	1.02	0.27	1.02	<0.01	0.90
5 days before admission	0.77	1.00	0.19	1.01	0.73	1.99	0.41	1.01	0.50	1.01	0.08	0.94
7 days before admission	0.48	1.01	0.046	1.02	0.40	1.02	0.09	1.03	0.40	1.01	<0.01	0.90
14 days before admission	0.13	1.02	<0.01	1.04	0.25	1.03	0.02	1.05	0.06	1.04	<0.01	0.89
30 days before admission	0.02	1.04	0.03	1.04	0.13	1.04	0.04	1.05	0.31	1.02	<0.01	0.87
40 days before admission	0.02	1.04	0.26	1.02	0.62	1.02	0.34	1.02	0.60	1.01	<0.01	0.87
50 days before admission	0.004	1.04	0.08	1.04	0.22	1.04	0.10	1.04	0.48	1.02	<0.01	0.82
60 days before admission	<0.01	1.05	0.13	1.03	0.32	1.03	0.16	1.04	0.44	1.02	<0.01	0.82
90 days before admission	<0.01	1.05	0.15	1.04	0.55	1.02	0.25	1.03	0.39	1.02	<0.01	0.78
IQR-P (μg/m^3)	12.8		11.08		15.40		31.79		0.62		31.80	

RR- relative risk calculated *per* 1 interquartile range of pollutant concentration increase (IQR-P)

association does not concern bronchitis. The authors suggest that a chronic character of bronchitis symptoms is a reason for postponing the visit to the hospital (Cody et al. 1992). Contrary to those results, a Spanish study has demonstrated a positive association between the O_3 content and hospital emergency admissions for COPD. This association has been substantiated for smaller changes of O_3 than those considered by us, yet the mean O_3 content was greater, especially during the warm season (Tenías et al. 2002). In the Air Pollution and Health: a European approach (APHEA) study, a widespread positive risk for COPD admissions has been demonstrated with the elevation of O_3 content in 6 European cities, each with different meteorological conditions and health services. The calculated relative risks have been consistent for 2 days' exposure/exacerbation lags, although averages up to 8 h have been taken into account (Anderson et al. 1997). The present findings on O_3 are at variance with the literature data. The divergent results may be explicable by the nature of O_3 as a pollutant with a delayed secondary reactivity. One-day averages, considered in the above-quoted studies could not exactly mirror the actual change in O_3 content, but rather be secondary to the interaction with other reactive pollutants.

In the afore-mentioned APHEA study, NO_2 content has been associated with increased risk for COPD-related admissions, apart from the data from Amsterdam, for up to 2 days' lag. In comparison to our present data, NO_2 rise connected with a significant change in admission rate has been substantially greater (50 μg/m^3 *versus* 15.5 μg/m^3 in our dataset) (Anderson et al. 1997). Indirect data for NO_2 are available in a study from Brussels analyzing COPD medication costs. That study shows an increase in COPD medication costs starting off with a 4 μg/m^3 NO_2 rise. Although that does not directly confirm the impact of NO_2 rise on COPD patients, it speaks for a likely connection with symptoms (Simons et al. 2016). On the other hand, a review of 5 cohorts from the European Study of Cohorts for Air Pollution Effects has failed to show any significant associations between NO_2 and NOx, and the chronic bronchitis symptoms. That study, however, differs in terms of the exposure measurement method as it employs the 14-day periods (Cai et al. 2016). In contrast, another study, which considers daily averages of NO_2 content, shows that NO_2 elevation is consistently associated with the COPD hospitalization risk. Further, the association is stronger than that for particulate pollutants. The authors emphasize that taking a longer exposure/exacerbation lag into consideration considerably increases the calculated risk of hospitalization (Faustini et al. 2013).

In conclusion, this study employed a new approach in the evaluation of hospitalization risk due to COPD-related respiratory exacerbations after exposure to gaseous pollutants, taking into account the impact of a co-existing cardiovascular disease. The study demonstrates that a co-existing cardiovascular malady, although not a direct underlier of respiratory complaints, enhanced the hospitalization risk due to COPD exacerbations, particularly when the pollutant SO_2 and NOx contents rose. In our opinion, cardiovascular co-morbidity ought to be taken into account in the true assessment of risks of respiratory exacerbations in COPD patients due to exposure to gaseous pollutants.

Conflicts of Interest The authors declare no conflicts of interest in relation to this article.

References

Anderson HR, Spix C, Medina S, Schouten JP, Castellsague J, Rossi G, Zmirou D, Touloumi G, Wojtyniak B, Ponka A, Bacharova L, Schwartz J, Katsouyanni K (1997) Air pollution and daily admissions for chronic obstructive pulmonary disease in 6 European cities: results from the APHEA project. Eur Respir J 10:1064–1071

Arbex MA, de Souza Conceição GM, Cendon SP, Arbex FF, Lopes AC, Moysés EP, Santiago SL, Saldiva PH, Pereira LA, Braga AL (2009) Urban air pollution and chronic obstructive pulmonary disease-related emergency department visits. J Epidemiol Community Health 63:777–783

Cai Y, Schikowski T, Adam M et al (2016) Cross-sectional associations between air pollution and chronic bronchitis: an ESCAPE meta-analysis across five cohorts. Thorax 69(11):1005–1014

Cody RP, Weisel CP, Birnbaum G, Lioy PJ (1992) The effect of ozone associated with summertime photochemical smog on the frequency of asthma visits to hospital emergency departments. Environ Res 58:184–194

de Hartog JJ, Hoek G, Peters A, Timonen KL, Ibald-Mulli A, Brunekreef B, Heinrich J, Tiittanen P, van Wijnen JH, Kreyling W, Kulmala M, Pekkanen J (2003) Effects of fine and ultrafine particles on cardiorespiratory symptoms in elderly subjects with coronary heart disease: the ULTRA study. Am J Epidemiol 157:613–623

Donaldson G, Seemungal T, Bhowmik A, Wedzicha J (2002) Relationship between exacerbation frequency and lung function decline in chronic obstructive pulmonary disease. Thorax 57(10):847–852

Faustini A, Stafoggia M, Colais P, Berti G, Bisanti L, Cadum E, Cernigliaro A, Mallone S, Scarnato C, Forastiere F, EpiAir Collaborative Group (2013) Air pollution and multiple acute respiratory outcomes. Eur Respir J 42(2):304–313

Hawkins NM, Petrie MC, Jhund PS, Chalmers GW, Dunn FG, McMurray JJ (2009) Heart failure and chronic obstructive pulmonary disease: diagnostic pitfalls and epidemiology. Eur J Heart Fail 11(2):130–139

Kitzman DW, Gardin JM, Gottdiener JS, Arnold A, Boineau R, Aurigemma G, Marino EK, Lyles M, Cushman M, Enright PL, Cardiovascular Health Study Research Group (2001) Importance of heart failure withpreserved systolic function in patients 65 years of age. CHS ResearchGroup. Cardiovascular Health Study. Am J Cardiol 87:413–419

Ko FW, Chan KP, Hui DS, Goddard JR, Shaw JG, Reid DW, Yang IA (2016) Acute exacerbation of COPD. Respirology 21(7):1152–1165

Lagorio S, Forastiere F, Pistelli R, Iavarone I, Michelozzi P, Fano V, Marconi A, Ziemacki G, Ostro BD (2006) Air pollution and lung function among susceptible adult subjects: a panel study. Environ Health 5:11

Peacock JL, Anderson HR, Bremner SA, Marston L, Seemungal TA, Strachan DP, Wedzicha JA (2011) Outdoor air pollution and respiratory health in patients with COPD. Thorax 66:591–596

Penttinen P, Timonen KL, Tiittanen P, Mirme A, Ruuskanen J, Pekkanen J (2001) Ultrafine particles in urban air and respiratory health among adult asthmatics. Eur Respir J 17:428–435

Rodriguez C, Tonkin R, Heyworth J, Kusel M, De Klerk N, Sly PD, Franklin P, Runnion T, Blockley A, Landau L, Hinwood AL (2007) The relationship between outdoor air quality and respiratory symptoms in young children. Int J Environ Health Res 17:351–360

Rushton CA, Satchithananda DK, Jones PW, Kadam UT (2015) Non-cardiovascular comorbidity, severity and prognosis in non-selected heart failure populations: a systematic review and meta-analysis. Int J Cardiol 196:98–106

Shah AS, Langrish JP, Nair H, McAllister DA, Hunter AL, Donaldson K, Newby DE, Mills NL (2013) Global association of air pollution and heart failure: a systematic review and meta-analysis. Lancet 382 (9897):1039–1048

Simons K, Devos S, Putman K, Coomans D, Van Nieuwenhuyse A, Buyl R (2016) Direct cost saving potential in medication costs due to a reduction inoutdoor air pollution for the Brussels Capital Region. Sci Total Environ 562:760–765

Stieb DM, Szyszkowicz M, Rowe BH, Leech JA (2009) Air pollution and emergency department visits for cardiac andrespiratory conditions: a multi-city time-series analysis. Environ Health 8:25

Tenías JM, Ballester F, Pérez-Hoyos S, Rivera ML (2002) Air pollution and hospital emergency room admissions for chronic obstructive pulmonary disease in Valencia, Spain. Arch Environ Health 57:41–47

Woodhead M, Blasi F, Ewig S, Garau J, Huchon G, Ieven M, Ortqvist A, Schaberg T, Torres A, van der Heijden G, Read R, Verheij TJ, Joint Taskforce of the European Respiratory Society and European Society for Clinical Microbiology and Infectious Diseases (2011) Guidelines for the management of adult lower respiratory tract infections–full version. Clin Microbiol Infect 17(Suppl 6):E1–E59

Advs Exp. Medicine, Biology - Neuroscience and Respiration (2018) 41: 19–29
https://doi.org/10.1007/5584_2018_222

Published online: 9 June 2018

Chemotherapy-Induced Takotsubo Syndrome

Monika Budnik, Jakub Kucharz, Paweł Wiechno, Tomasz Demkow, Janusz Kochanowski, Elżbieta Górska, and Grzegorz Opolski

Abstract

Cardiovascular complications are a significant problem in systemically treated cancer patients. One such complication is Takotsubo cardiomyopathy, also known as Takotsubo syndrome. It is most frequently defined as a sudden and transient left or right ventricular systolic dysfunction; mimicking acute coronary syndrome, but without the associated changes in coronary arteries. Takotsubo syndrome is a relatively little known complication that appears in the course of oncological treatment, and its incidence has not yet been established. In this study, we reviewed Medline database according to case reports concerning takotsubo syndrome appearing after systemic treatment in oncological patients. We took into consideration all types of anticancer drugs. We reviewed the changes reported in the electrocardiography, echocardiography, and coronary angiography, and also the level of troponin, a marker of acute coronary syndrome elevation. In view of the increasing frequency of cardiac complications reported in patients receiving systemic oncological treatment, Takotsubo syndrome appears to be underdiagnosed. However, the syndrome may be linked to potentially fatal complications such as cardiogenic shock or cardiac arrest. Therefore, it seems essential to carry out appropriate diagnostic procedures for every patient experiencing clinical side effects of oncopharmacotherapy. In patients with chest pain and dyspnea during or after treatment, Takotsubo syndrome should be considered, particularly that the syndrome requires a different therapy approach than that used in a coronary

M. Budnik, J. Kochanowski, and G. Opolski
First Department of Cardiology, Warsaw Medical University, Warsaw, Poland

J. Kucharz (✉), P. Wiechno, and T. Demkow
Department of Uro–Oncology, Maria Sklodowska-Curie Memorial Cancer Center and Institute of Oncology, Warsaw, Poland
e-mail: jakub.kucharz@gmail.com

E. Górska
Department of Laboratory Diagnostics and Clinical and Clinical Immunology of Developmental Age, Warsaw Medical University, Warsaw, Poland

syndrome. Diagnostic procedures should include echocardiogram and the assessment of myocardial necrosis markers and natriuretic peptides.

Keywords
Cancer • Cardiomyopathy • Cardiotoxicity • Chemotherapy • Coronary syndrome • Takotsubo syndrome

1 Introduction

Cardiovascular complications are a significant issue in cancer patients. Such complications stem from a growing prevalence of various types of cancer in the general population, and the fact that treatment is administered to older patients and those with diseases that increase a cardiac risk, e.g., diabetes, obesity, metabolic syndrome, or chronic kidney disease (Zamorano et al. 2016; Napora et al. 2013). Cardiovascular complications are also associated with oncological treatments themselves, as the treatments have become more aggressive over the last decade and now often combine various therapeutic modes. In addition, a number of new drugs have been introduced into clinical practice, many of which have significant cardiotoxicity that results, inter alia, in a decreased left ventricular ejection fraction. Conservative therapy of the early-stage breast cancer is one example of an intensive combination oncological therapy that carries a significant risk of cardiovascular complications. For instance, in patients with luminal B hormone-receptor positive (estrogen-receptor and/or progesterone (HER2)-receptor positive) breast cancer, treatment begins with surgery, which is followed by anthracycline-based chemotherapy and then by the administration of taxanes and trastuzumab (Senkus et al. 2015). Depending on individual medical circumstances, radiotherapy can be the next treatment stage. After administration of cytostatics, treatment also includes endocrine therapy, which includes either tamoxifen or aromatase inhibitors, depending on the patient's menopausal status. Such patient is exposed to the toxic effects of chemotherapy, radiotherapy, targeted treatment, and hormonal manipulations, all of which become cumulative. Anthracyclines lead to anthracycline-induced congestive heart failure, taxanes negatively affect the heart's electrical conduction system, trastuzumab can cause acute left ventricular dysfunction and conduction abnormalities, and radiotherapy is associated with more frequent and earlier onset of ischemic heart disease (Khanji et al. 2013), while endocrine therapy is associated with endothelial dysfunction, thromboembolic episodes, and, when aromatase inhibitors are used, with more frequent ischemic heart disease, strokes, and ventricular dysfunction (Zamorano et al. 2016).

Other commonly used drugs with potential cardiotoxic effects are the derivatives of fluoropyrimidine, such as 5-fluorouracil and its precursor, capecitabine. These drugs are widely used in gastrointestinal malignancies, e.g., colorectal, gastric, pancreatic, and esophageal cancers, breast cancers, and head and neck tumors (Knott et al. 2014; Kobayashi et al. 2009). Typically, cardiovascular toxicity of these drugs appears as acute ischemic heart disease, which is probably associated with a coronary vasospasm, often appearing during extended intravenous infusions that may last up to 46 h (Zamorano et al. 2016).

5–Fluorouracil- or capecitabine-induced cardiotoxicity affects 1–18% of patients undergoing the treatment (Sorrentino et al. 2012; Stewart et al. 2010). Cardiotoxicity appears more frequently when 5-fluorouracil is administered through extended intravenous infusions than when it is taken orally. Symptoms of cardiotoxicity reappear in approximately half of the patients who take the drug a second time (Saif et al. 2009). The cause of this cardiotoxicity remains unclear, but the most frequently proposed mechanism of action is a coronary vasospasm, leading to an acute ischemic heart episode (Ozturk et al. 2013; Grunwald et al. 2012).

Cardiovascular complications after administration of capecitabine are less frequent and include anginal chest pain with shortness of breath, myocardial infarction, cardiogenic shock, and sudden cardiac death. It seems these complications appear less frequently with capecitabine than with 5-fluorouracil. The pathogenesis of capecitabine cardiotoxicity is unclear. The proposed mechanisms include coronary vasospasm and a direct toxic effect on cardiomyocytes (Qasem et al. 2016; Klag et al. 2014). Trastuzumab causes left ventricular dysfunction in about 2–25% of patients, and the most likely mechanism is the interference with the proper development, repair, and survival of cardiomyocytes (Yeh and Bickford 2009). The underlying mechanisms of cardiotoxic effects of bevacizumab, an inhibitor of the vascular endothelial growth factor (VEGF), are unclear. Experimental studies performed in mice treated with bevacizumab demonstrate a high content of pro-inflammatory cytokines and chemoattractant proteins in the plasma and cardiac muscle tissue, which may be responsible for cardiotoxicity (Drímal et al. 2006). VEGF inhibition is associated with increased frequency of embolic events. A meta-analysis has shown that the risk of embolic complications is higher in cancer patients who receive bevacizumab than in those who do not (Scappaticci et al. 2007). The frequency of embolic complications, like myocardial infarction and angina pectoris, is set at 4.4% (Franco et al. 2008).

Differential diagnosis of a cardiovascular disease in patients undergoing oncological treatment can be challenging because of the possible co-existence of the following: (1) cardiotoxic effects of the treatment itself; (2) the patient's pre-existing health problems, e.g., coronary atherosclerosis, anemia induced either by the cancer or by systemic cancer treatment; (3) effects of earlier radiotherapy; and (4) effects of treatment with either vinca alkaloids or taxenes, which can produce autonomic neuropathy symptoms.

Takotsubo syndrome (TTS) is a relatively little known complication that appears in the course of oncological treatment and its incidence has not yet been established. TTS, also known as apical ballooning syndrome or stress cardiomyopathy, is most frequently defined as a sudden, transient, and reversible left ventricular (LV) systolic dysfunction (Goel et al. 2014; Kim et al. 2008; Khakoo et al. 2005). TTS's symptoms are similar to those of acute coronary syndrome, but without the associated changes in coronary arteries (Gianni et al. 2006; Bybee et al. 2004). The term "takotsubo" cardiomyopathy was first coined in 1990. The name comes from the echocardiography or ventriculography observation that the LV assumes a specific shape in the late systolic phase, which is reminiscent of a takotsubo, a traditional Japanese fishing trap, used for catching octopi, with a flat bottom and a narrow neck (Fig. 1).

Most TTS patients are postmenopausal women who have chest pain, increased levels of myocardium necrosis markers, high concentration of natriuretic peptides, and ischemic changes in the electrocardiogram (ECG) (Budnik et al. 2016; Budnik et al. 2015). Symptoms are typically preceded by physical or emotional stress, but may also be associated with exacerbations of chronic diseases or be complications of cancer treatment (Coli et al. 2015; Baumann et al. 2014). The most likely underlying mechanism of TTS is an excess of circulating catecholamines. Diagnostic criteria for TTS of the Heart Failure Association of the European Society of Cardiology are summarized in Table 1 (Lyon et al. 2016).

In cancer patients, TTS can result from significant emotional stress associated with receiving a cancer diagnosis. It can also appear in the course of paraneoplastic syndrome (Finsterer et al. 2015), pain syndromes (Singh and Harle 2014), and as a complication of surgical treatment (Burgy et al. 2014) or chemotherapy (Malley and Watson 2016) (Table 2).

2 Diagnosis

Basic procedures necessary for TTS diagnosis include an analysis of myocardium necrosis markers, ECG, echocardiogram, and angiography. Rarely, in patients who are in stable condition and who come to the hospital more than 48 h after onset of symptoms, a CT scan of coronary arteries can also be considered (Lyon et al. 2016).

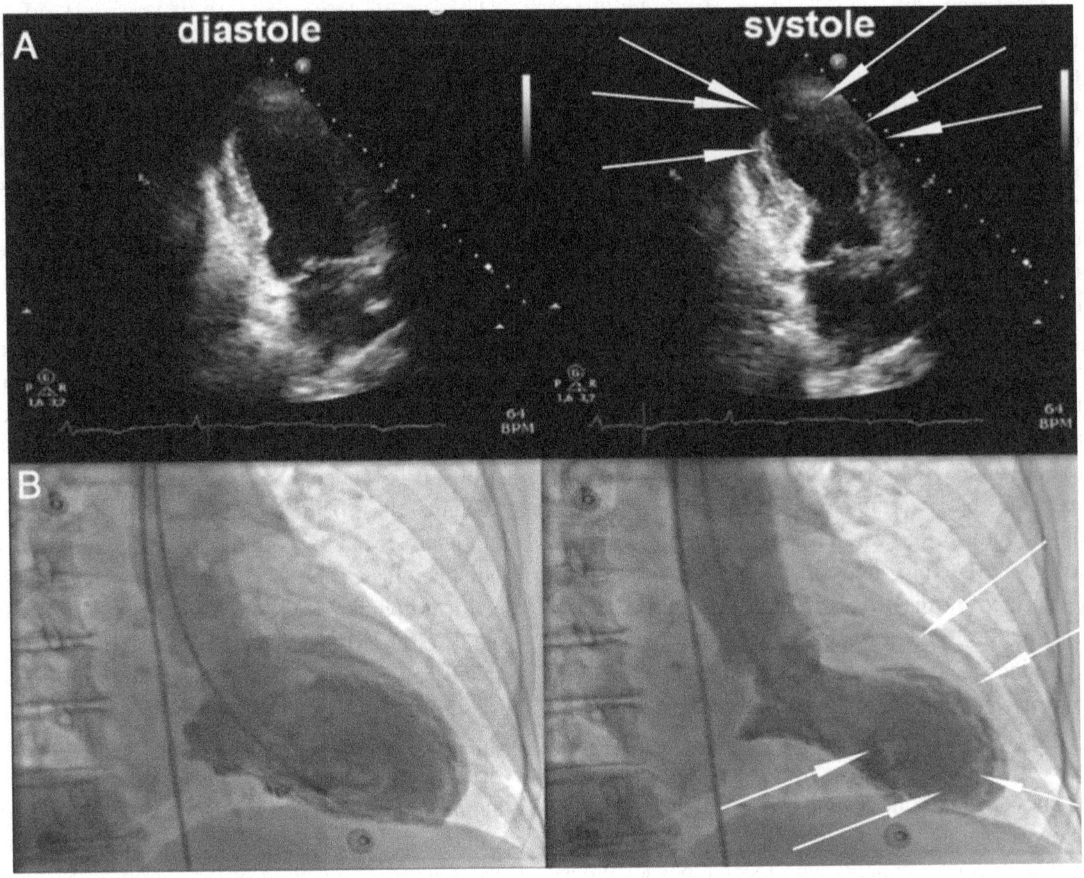

Fig. 1 Takotsubo syndrome in the echocardiography image (**a**) and ventriculography image (**b**). Arrows show dyskinesis of the left ventricular apex

Table 1 Diagnostic criteria for takotsubo syndrome according to the Heart Failure Association of the European Society of Cardiology

1	Transient regional wall motion abnormalities of LV or RV myocardium which are frequently, but not always, preceded by a stressful emotional or physical trigger
2	Regional wall motion abnormalities extending beyond a single epicardial vascular distribution, which result in circumferential dysfunction of ventricular segments
3	Absence of atherosclerotic coronary artery disease, acute plaque rupture, thrombus formation, coronary dissection, or other pathological conditions that could explain temporary LV dysfunction (e.g., hypertrophic cardiomyopathy or viral myocarditis)
4	New and reversible electrocardiography (ECG) abnormalities (ST-segment elevation, ST depression, LBBB, T-wave inversion, and/or QT prolongation) during acute phase (preceding 3 months)
5	Significantly elevated serum natriuretic peptide (BNP or NT-proBNP) during acute phase
6	Positive, although relatively small, elevation in cardiac troponin measured with a conventional assay, i.e., disparity between the troponin level and the amount of dysfunctional myocardium involved
7	Recovery of ventricular systolic function on cardiac imaging at follow-up in 3–6 months

LV left ventricle, *RV* right ventricle, *LBBB* left bundle branch block, *BNP* brain natriuretic peptide, *NT-proBNP* N-terminal pro b-type natriuretic peptide

Table 2 Chemotherapy-induced takotsubo syndrome (TTS)

Chemotherapy	Malignancy	Age	Gender	Onset after treatment induction	Troponin	ECG abnormalities	Echo	Complication
5-FU Lim et al. (2013)	Rectal adenocarcinoma	66	F	3 weeks	TnI 0.23 mmol/L	Giant T wave inversions in all leads, except AVR and AVL	Apical akinesis; EF 30%	Apical Thrombus
5-FU Kobayashi et al. (2009)	Rectal adenocarcinoma	62	F	NA	TnT 0.82 ng/mL	ST segment elevation in leads V1-V3	Extensive apical akinesis	None
5-FU, oxaliplatin and calcium folinate Basselin et al. (2011)	Colon adenocarcinoma	48	M	24 h	0.5 ng/mL	Apico-lateral ST segment and T wave depression	Hypokinesia in all apical and median segments; EF 15%	Cardiogenic shock
mFOLFOX-6 Sundaravel et al. (2017)	Rectal adenocarcinoma	35	M	1st day	NA	NA	Diffuse left ventricle hypokinesis; EF 19%	Cardiogenic shock
5-FU Stewart et al. (2010)	Colon adenocarcinoma	58	F	46 h	TnT 0.64 mg/L	T wave inversion	Hypokinesis in all apical and median segments	None
5-FU Knott et al. (2014)	Sigmoid colon adenocarcinoma	59	M	46 h	TnI 1.00 μg/L	ST segment elevation and hyperacute T-wave changes in septal, inferior, and lateral leads	Global left ventricular dysfunction; EF 10%	Cardiac arrest
5-FU Ozturk et al. (2013)	Gastric carcinoma	48	M	34 h	2.87 ng/mL	T wave inversion in leads V4–V5	Hypokinesis in mid-apical segments and hyperkinesis in basal segments	VF
5-FU Grunwald et al. (2012)	Colon carcinoma	60	F	26 h	TnI 0.38 ng/mL	T wave flattening in lateral precordial leads	Global hypokinesis of left ventricle with apical ballooning in systole and diastole; EF 15–20%	None
5-FU Singh and Harle (2014)	Colorectal carcinoma	79	F	2 weeks	TnT 1.06 mg/L	ST segment elevation in the anterior and inferior leads	Akinetic apical and periapical segments; EF 34%	None
Capecitabine Stewart et al. (2010)	Colon carcinoma	81	F	48 h	TnT 0.35 mg/L	T wave inversion in antero-septal, inferior, and lateral leads	Akinesis of distal halves of the anterior and inferior walls; EF 35%	None
Capecitabine Qasem et al. (2016)	Breast carcinoma	47	M	48 h	TnI 0.19 ng/mL	Nonspecific ST-T wave changes	Antero-apical wall motion abnormality; EF 35%	None
Capecitabine Y-Hassan et al. (2013)	Caecum adenocarcinoma	55	M	28 h	hsTnT 89 ng/L	ST elevations in leads V2–V6, II, aVR, and aVF	Global hypokinesia of LV; EF 15–20%	Cardiogenic shock

(continued)

Table 2 (continued)

Chemotherapy	Malignancy	Age	Gender	Onset after treatment induction	Troponin	ECG abnormalities	Echo	Complication
Capecitabine Klag et al. (2014)	Breast carcinoma	81	F	6 h	hsTnT 282 pg/mL	ST elevation in leads V2–V4, high T waves in leads II, III, AVF, ST depression in AVL	Apical dyskinesia with typical ballooning and diminished systolic function; EF 53%	None
Trastuzumab Khanji et al. (2013)	Breast carcinoma	50	F	30 min after 11th infusion	TnT 0.15 pg/mL	T-wave inversion in leads I, aVL and V1–V3.	Systolic bulging at mid-LV level with normal apical and basal contractions	None
Bevacizumab Franco et al. (2008)	Colon carcinoma	76	M	2 days	Tn 11.4 ng/mL	Anterior and inferior ST segment elevations	Decreased systolic function with apical akinesis and a hyperdynamic basal segment of LV	None
Bevacizumab Franco et al. (2008)	Metastatic non-small cell lung carcinoma	61	M	3 weeks	Tn 2.5 ng/mL	Fresh Q waves in inferior and precordial leads, diffuse ST segment elevations	On cardiac catheterization-severely depressed LV function	None
Rituximab Ng et al. (2015)	CLL, stage C	66	M	40 min	Tn 147 ng/L	ST segment elevation in leads I, II, and V4–V6	Hypokinesis of anterior wall	None
Cetuximab Kim et al. (2008)	Laryngeal squamous cell carcinoma, stage III	67	M	3 days	TnI 0.96 ng/mL	ST segment elevation in leads V2–V6	Impaired LV function with EF 20%. Severe hypokinesis of distal anterior and apical regions.	Cardiac arrest, cardiogenic shock
Axitinib Ovadia et al. (2015)	Renal cell carcinoma	71	F	24 h	TnI 6.95 µg/L	Antero-lateral ST elevation	Global hypokinesis of LV with anterior apical ballooning; EF 20–25%	None
Cytarabine Baumann et al. (2014)	AML	58	M	After the 2nd course of chemotherapy	NA	T wave inversions in leads II, III, aVF, and V3–V6	Dyskinesis of mid and apical portions of both ventricles; EF 20%	Cardiogenic shock
Cytarabine and daunorubicin Goel et al. (2014)	AML	55	M	6 days	TnI 38.64 ng/mL	ST segment elevation in leads I, aVL, V5–V6	Mild anterior, septal, apical, inferior, and lateral wall hypokinesis; EF 30–35%	None
Interleukin 2 Damodaran et al. (2014)	Melanoma	55	F	Before course 2, cycle 1 of chemotherapy	0	T wave inversions in anterolateral leads	Hypokinesis of apico-lateral wall segments; EF 45%	None

Interferon alpha-2b Khakoo et al. (2005)	Melanoma	56	M	8 months	NA	T wave flattening in leads V2-V5	Severe global systolic dysfunction; EF 15%	Cardiogenic shock
Oxaliplatin Coli et al. (2015)	Colorectal carcinoma	67	F	At the end of 3rd course of chemotherapy	TnI 1.10 ng/mL	ST segment elevation w V1, aVL, T-wave inversion, and QTc elongation	Akinesis of all mid-segments, with sparing of basal and apical segments	None
Lomustine, cytarabine, cyclophosphomide and etoposide Malley and Watson (2016)	DLBCL, stage IV B	73	F	7 days after chemotherapy	hsTn 400 ng/l	T wave inversion in leads II, aVF, and V3 – V6	Mildly impaired systolic function and antero-septal hypokinesis; EF 45%.	None

ECG electrocardiography, *F* female, *M* male, *5-FU* 5–fluorouracil, *TnI* troponin I, *TnT* troponin T, *hsTnT* high-sensitive troponin T, *EF* ejection fraction, *NA* not available, *mFOLFOX* modified FOLFOX (folinic acid, 5-fluorouracil, oxaliplatin), *LV* left ventricle, *VF* ventricular fibrillation, *CLL* chronic lymphatic leukemia, *AML* acute myeloid leukemia, *DLBCL* diffuse large B cell lymphoma

Such diagnostic procedures should be done for all patients with cardiovascular problems that become manifest after chemotherapy. The procedures are particularly important in patients treated with fluoropyrimidines who experience retrosternal pain; the exemplary is pain typically associated with transient aliments appearing after infusion of 5-fluorouracil (Zamorano et al. 2016).

Invasive diagnostic procedures, such as angiography, can be performed safely in patients undergoing oncological treatment. It is necessary, however, to monitor peripheral blood morphology to rule out thrombocytopenia associated with systemic treatment. If the platelet count is greater than 50,000/mm^3, most treatment procedures can be carried out safely.

3 Treatment

Experts advise to carry out a risk stratification once TTS is diagnosed (Table 3). If the patient is found to have a high risk of complications, i.e., when there is at least one significant or two less significant risk factors, he should be monitored for at least 72 h. If the ejection fraction is less than 45%, the patient should be given a beta-adrenolytic and an angiotensin-converting-enzyme inhibitor. A beta-adrenolytic should also be used in patients who experience atrial fibrillation, atrial tachycardia, ventricular tachycardia, or ventricular fibrillation. If a thrombus forms in the apex, low-molecular-weight heparin should be administered, followed by 3 months of oral anticoagulant therapy. Even when there is no thrombus, low-molecular-weight heparin should be consider prophylactically until wall motion

Table 3 Risk stratification in takotsubo syndrome (TTS) according to the Heart Failure Association of the European Society of Cardiology (Lyon et al. 2016)

Risk factor	Higher risk	Lower risk
Age (years)	70–75	< 70
ECG		
QTc (ms)	≥ 500	< 500
Pathological Q wave	Yes	No
Persistent ST segment elevation	Yes	No
EF (%)	35–45	≥ 45
Physical stress	Yes	No
Natriuretic peptides	Yes	No
BNP (pg/mL)	≥ 600	< 600
NTproBNP (pg/mL)	≥ 2000	< 2000
Bystander obstructive CAD	Yes	No
Biventricular involvement	Yes	No
Major risk factors		
Age (years)	≥ 75	< 70
Systolic BP (mmHg)	< 110	≥ 110
Clinical pulmonary edema	Yes	No
Unexplained syncope, VT, or VF	Yes	No
EF (%)	<35	≥45
LVOTO (mmHg)	≥40	Absent or <40
Significant mitral regurgitation	Yes	No
Apical thrombus	Yes	No
VSD or LV rupture	Yes	No

BP blood pressure, *CAD* coronary artery disease, *NTproBNP* recombinant N-terminal proBNP protein, *BNP* brain natriuretic peptide, *EF* ejection fraction, *QTc* QT interval with a correction based on the patient's heart rate, *ECG* electrocardiogram, *VT* ventricular tachycardia, *VF* ventricular fibrillation, *LVOTO* left ventricular outflow tract obstruction, *VSD* ventricular deptal defect, *LV* left ventricle abnormalities subside. If cardiogenic shock develops, it is essential to halt chemotherapy that caused the symptoms and to administer levosimendan, and, in severe cases, to offer the patient extracorporeal membrane oxygenation or a left ventricular assist device (Lyon et al. 2016). TTS can be relapsing if there is recurrent exposure to its causal factors (Opolski et al. 2016). Patients receiving systemic treatment should be seen as being at risk of TTS recurrence if the syndrome was initially induced by administration of a drug. There are currently no recommendations about procedures that could prevent TTS relapse. Initially, the elevation of circulating catecholamines associated with TTS suggested that beta-adrenolytic could prevent relapse, but this has not been confirmed in clinical trials (Templin et al. 2015). Administration of angiotensin converting enzyme inhibitors could potentially reduce the frequency of TTS relapses (Brunetti et al. 2016).

4 Complications

Patient prognosis is better in TTS than in myocardial infarction. Unfortunately, complications sometimes appear, including right ventricular dysfunction, cardiac rhythm abnormalities (atrial fibrillation or bradycardia), asystolic cardiac arrest and atrial-fibrillation-induced cardiac arrest, cardiogenic shock, and rupture of the left ventricular wall. Hospital mortality ranges from 1% to 4.5%, while mortality 5 years after TTS ranges from 5% to 22% (Lyon et al. 2016). There is a high frequency of severe complications and fatalities among patients with post-chemotherapy TTS. This could result from insufficient diagnosis in patients who do not present severe clinical symptoms, or from the overlap of various mechanisms of cardiotoxicity.

5 Conclusions

Because of the increasing frequency of cardiac complications in patients receiving oncological treatment, it is necessary to carry out appropriate diagnostic procedures for every patient experiencing clinical side effects. Differential diagnosis should take into account not only possible side effects of oncological drugs and cardiovascular complications associated with cancer, e.g. pulmonary embolism, but also relatively rare disorders, like TTS, which, when occurring in cancer patients, can be associated with either stress or with the action of drugs used in systemic treatment. It appears that TTS is underdiagnosed in oncology departments. It is usually found only when patients experience complications, such as cardiogenic shock or cardiac arrest. Differential diagnosis for every patient experiencing chest pain and difficulty breathing should take TTS into account; diagnostics should include the echocardiogram, assessment of myocardium necrosis markers, and determination of natriuretic peptides levels.

Conflicts of Interest The authors declare no conflicts of interest in relation to this article.

References

Basselin C, Fontanges T, Descotes J, Chevalier P, Bui-Xuan B, Feinard G, Timour Q (2011) 5-Fluorouracil-induced Tako-Tsubo-like syndrome. Pharmacotherapy 31:226

Baumann S, Huseynov A, Goranova D, Faust M, Behnes M, Nolte F, Heidenreich D, Hofmann WK, Borggrefe M, Akin I, Klein S (2014) Takotsubo cardiomyopathy after systemic consolidation therapy with high-dose intravenous cytarabine in a patient with acute myeloid leukemia. Oncol Res Treat 37:487–490

Brunetti ND, Santoro F, De Gennaro L, Correale M, Gaglione A, Di Biase M (2016) Drug treatment rates with beta-blockers and ACE-inhibitors/angiotensin receptor blockers and recurrences in takotsubo cardiomyopathy: a meta-regression analysis. Int J Cardiol 214:340–342

Budnik M, Piatkowski R, Kochanowski J, Glowczynska R, Gorko D, Kowalik R, Pietrasik A, Opolski G (2015) The oldest patient with takotsubo cardiomyopathy. J Geriatr Cardiol 12:588–589

Budnik M, Kochanowski J, Piatkowski R, Wojtera K, Peller M, Gaska M, Glowacka P, Karolczak P, Ochijewicz D, Opolski G (2016) Simple markers can distinguish Takotsubo cardiomyopathy from ST segment elevation myocardial infarction. Int J Cardiol 219:417–420

Burgy M, Brossat H, Barthelemy P, Imperiale A, Trinh A, Hazam CA, Bergerat JP, Mathelin C (2014) First report of trastuzumab treatment after postoperative takotsubo cardiomyopathy. Anticancer Res 34:3579–3582

Bybee KA, Kara T, Prasad A, Lerman A, Barsness GW, Wright RS, Rihal CS (2004) Systematic review: transient left ventricular apical ballooning: a syndrome that mimics ST-segment elevation myocardial infarction. Ann Intern Med 141:858–865

Coli S, Pigazzani F, Gaibazzi N (2015) Midventricular takotsubo cardiomyopathy after oxaliplatin infusion: an unreported side effect. J Cardiovasc Med 16:646–649

Damodaran S, Mrozek E, Liebner D, Kendra K (2014) Focal takotsubo cardiomyopathy with high-dose interleukin-2 therapy for malignant melanoma. J Natl Compr Cancer Netw 12(12):1666–1670

Drímal J, Zúrová-Nedelcevová J, Knezl V, Sotníková R, Navarová J (2006) Cardiovascular toxicity of the first line cancer chemotherapeutic agents: doxorubicin, cyclophosphamide, streptozotocin and bevacizumab. Neuro Endocrinol Lett 27(Suppl 2):176–179

Finsterer J, Stöllberger C, Pulgram T (2015) Paraneoplastic takotsubo syndrome with ventricular thrombus and stroke. Herz 40:632–634

Franco TH, Khan A, Joshi V, Thomas B (2008) Takotsubo cardiomyopathy in two men receiving bevacizumab for metastatic cancer. Ther Clin Risk Manag 4:1367–1370

Gianni M, Dentali F, Grandi AM, Sumner G, Hiralal R, Lonn E (2006) Apical ballooning syndrome or takotsubo cardiomyopathy: a systematic review. Eur Heart J 27:1523–1529

Goel S, Sharma A, Garg A, Chandra A, Shetty V (2014) Chemotherapy induced takotsubo cardiomyopathy. World J Clin Cases 2:565–568

Grunwald MR, Howie L, Diaz LA (2012) Takotsubo cardiomyopathy and fluorouracil: case report and review of the literature. J Clin Oncol 30:e11–e14

Khakoo AY, Halushka MK, Rame JE, Rodriguez ER, Kasper EK, Judge DP (2005) Reversible cardiomyopathy caused by administration of interferon alpha. Nat Clin Pract Cardiovasc Med 2:53–57

Khanji M, Nolan S, Gwynne S, Pudney D, Ionescu A (2013) Tako-Tsubo syndrome after trastuzumab – an unusual complication of chemotherapy for breast cancer. Clin Oncol (R Coll Radiol) 25:329

Kim L, Karas M, Wong SC (2008) Chemotherapy-induced takotsubo cardiomyopathy. J Invasive Cardiol 20:338–340

Klag T, Cantara G, Ong P, Kaufmann M, Sechtem U, Athanasiadis A (2014) Epicardial coronary artery spasm as cause of capecitabine-induced takotsubo cardiomyopathy. Clin Res Cardiol 103:247–250

Knott K, Starling N, Rasheed S, Foran J, Cafferkey C, Rosen S, Nicholson A, Baksi J, Lyon A (2014) A case of Takotsubo syndrome following 5-fluorouracil chemotherapy. Int J Cardiol 177:e65–e67

Kobayashi N, Hata N, Yokoyama S, Shinada T, Shirakabe A, Mizuno K (2009) A case of Takotsubo cardiomyopathy during 5-fluorouracil treatment for rectal adenocarcinoma. J Nippon Med Sch 76:27–33

Lim SH, Wilson SM, Hunter A, Hill J, Beale P (2013) Takotsubo cardiomyopathy and 5-fluorouracil: getting to the heart of the matter. Case Rep Oncol Med 2013:206765

Lyon AR, Bossone E, Schneider B, Sechtem U, Citro R, Underwood SR, Sheppard MN, Figtree GA, Parodi G, Akashi YJ, Ruschitzka F, Filippatos G, Mebazaa A, Omerovic E (2016) Current state of knowledge on takotsubo syndrome: a position statement from the taskforce on Takotsubo syndrome of the heart failure Association of the European Society of Cardiology. Eur J Heart Fail 18:8–27

Malley T, Watson E (2016) A case of takotsubo cardiomyopathy after chemotherapy. Oxf Med Case Rep 2016:55–58

Napora M, Ganowicz E, Gorska R (2013) Correlation between the state of periodontal tissues and selected cardiovascular parameters in patients with type 2 diabetes. Centr Eur J Immunol 38(4):556–561

Ng KH, Dearden C, Gruber P (2015) Rituximab-induced Takotsubo syndrome: more cardiotoxic than it appears? BMJ Case Rep 2015:bcr2014208203. https://doi.org/10.1136/bcr-2014-208203

Opolski G, Budnik M, Kochanowski J, Kowalik R, Piatkowski R, Kochman J (2016) Four episodes of takotsubo cardiomyopathy in one patient. Int J Cardiol 203:53–54

Ovadia D, Esquenazi Y, Bucay M, Bachier CR (2015) Association between takotsubo cardiomyopathy and axitinib: case report and review of the literature. J Clin Oncol 33(1):e1–e3

Ozturk MA, Ozveren O, Cinar V, Erdik B, Oyan B (2013) Takotsubo syndrome: an underdiagnosed complication of 5–fluorouracil mimicking acute myocardial infarction. Blood Coagul Fibrinolysis 24:90–94

Qasem A, Bin Abdulhak AA, Aly A, Moormeier J (2016) Capecitabine-induced takotsubo cardiomyopathy: a case report and literature review. Am J Ther 23(5): e1188–e1192

Saif MW, Shah MM, Shah AR (2009) Fluoropyrimidine-associated cardiotoxicity: revisited. Expert Opin Drug Saf 8:191–202

Scappaticci FA, Skillings JR, Holden SN, Gerber HP, Miller K, Kabbinavar F, Bergsland E, Ngai J, Holmgren E, Wang J, Hurwitz H (2007) Arterial thromboembolic events in patients with metastatic carcinoma treated with chemotherapy and bevacizumab. J Natl Cancer Inst 99:1232–1239

Senkus E, Kyriakides S, Ohno S, Penault-Llorca F, Poortmans P, Rutgers E, Zackrisson S, Cardoso F; ESMO Guidelines Committee (2015) Primary breast cancer: ESMO clinical practice guidelines for diagnosis, treatment and follow-up Ann Oncol 26(suppl 5): v8–v30

Singh SB, Harle IA (2014) Takotsubo cardiomyopathy secondary in part to cancer-related pain crisis: a case report. J Pain Symptom Manag 48:137–142
Sorrentino MF, Kim J, Foderaro AE, Truesdell AG (2012) 5-Fluorouracil induced cardiotoxicity: review of the literature. Cardiol J 19:453–458
Stewart T, Pavlakis N, Ward M (2010) Cardiotoxicity with 5-fluorouracil and capecitabine: more than just vasospastic angina. Intern Med J 40:303–307
Sundaravel S, Alrifai A, Kabach M, Ghumman W (2017) FOLFOX induced takotsubo cardiomyopathy treated with impella assist device. Case Rep Cardiol 2017:8507096
Templin C, Ghadri JR, Diekmann J, Napp LC, Bataiosu DR, Jaguszewski M, Cammann VL et al (2015) Clinical features and outcomes of Takotsubo (stress) cardiomyopathy. N Engl J Med 373:929–938
Yeh ET, Bickford CL (2009) Cardiovascular complications of cancer therapy: incidence, pathogenesis, diagnosis, and management. J Am Coll Cardiol 53:2231–2247
Y-Hassan S, Tornvall P, Törnerud M, Henareh L (2013) Capecitabine caused cardiogenic shock through induction of global Takotsubo syndrome. Cardiovasc Revasc Med 14(1):57–61
Zamorano JL, Lancellotti P, Rodriguez Muñoz D, Aboyans V, Asteggiano R, Galderisi M, Habib G, Lenihan DJ, Lip GYH, Lyon AR, Lopez Fernandez T, Mohty D, Piepoli MF, Tamargo J, Torbicki A, Suter TM; ESC Scientific Document Group (2016) 2016 ESC position paper on cancer treatments and cardiovascular toxicity developed under the auspices of the ESC Committee for practice guidelines: The Task Force for cancer treatments and cardiovascular toxicity of the European Society of Cardiology (ESC). Eur Heart J 37:2768–2801

Advs Exp. Medicine, Biology - Neuroscience and Respiration (2018) 41: 31–39
https://doi.org/10.1007/5584_2018_203

Published online: 21 April 2018

Blood Group and Incidence of Asthma and Chronic Obstructive Pulmonary Disease

Bożena Mroczek, Zygmunt Sitko, Agnieszka Sujewicz, Weronika Wolińska, Iwona Karpeta-Pawlak, and Donata Kurpas

Abstract

Chronic respiratory diseases are determined by genetic predisposition, and environmental and socioeconomic variables. One genetic factor underlying susceptibility to such diseases can be the ABO blood group system. The purpose of this study was to investigate the hypothesis that there would be a relationship between the blood group and risk of developing asthma and chronic obstructive pulmonary disease (COPD). We reviewed medical history files of patients with the diagnosis of COPD or asthma, including those suffering from a cancerous disease accompanied by asthma or COPD, hospitalized from January 2016 to July 2017. The study involved 248 adult patients (85 women and 163 men; median age 65, range 20–95 years) with COPD (177; 71.4%) and asthma (71; 28.6%) with a known blood type. We found that the most frequent was blood type A (97; 39.1%) and the least blood type AB (26; 8.0%). The distribution of individual blood types in asthma was not different from that in COPD patients. Significant differences were found between the distribution of O and pooled non-O blood types (A, B, and AB); the pooled category was more frequent in both asthma and COPD patients. We conclude that the blood type is not a prognostic differentiating between the occurrence of asthma or COPD, although both are less common in people with blood type O. Further investigation is required to set the predisposing role of the ABO antigens for chronic respiratory diseases.

Keywords

Asthma · Blood group · Cancer · COPD · Genetic predisposition · Phenotype · Respiratory disease

B. Mroczek (✉), W. Wolińska, and I. Karpeta-Pawlak
Department of Humanities in Medicine, Pomeranian Medical University, Szczecin, Poland
e-mail: b_mroczek@data.pl

Z. Sitko
Voivodship Hospital in Szczecin Poland, Szczecin, Poland

A. Sujewicz
Department of Humanities in Medicine, Pomeranian Medical University, Szczecin, Poland

Department of General and Oncological Urology, Independent Public Hospitals Provincial Hospital, Szczecin, Poland

D. Kurpas
Department of Family Medicine, Wrocław Medical University, Wrocław, Poland

Opole Medical School, Opole, Poland

1 Introduction

Chronic obstructive pulmonary disease (COPD) and asthma are the most common chronic respiratory diseases; they involve inflammation within the lower airways, leading to bronchial obstruction (Kim and Rhee 2010). Respiratory diseases are the third leading cause of death in the EU. The annual mortality rate for COPD in Poland was 52.1 *per* 100,000 people in 2015 (Global Disease Burden 2013). According to WHO (2017), chronic obstructive pulmonary disease (COPD) affects approximately 210 million people throughout the world, of whom 3 million die each year. COPD has been the third most common cause of death in the world. Estimates show that 2 million people in Poland, 5.3% of the population, suffer from this condition, but as many as 80% of them have not been diagnosed. Some COPD cases are caused later in life by persistent asthma, which worsens health status and hinders the therapeutic process (Mathers and Loncar 2006). As opposed to COPD, asthma usually starts during the first years of life, but its severity does not substantially change over time (Castro-Rodriguez et al. 2000). The incidence of asthma in Poland is 1,250 patients *per* 10,000 people, which means that about 4.5 million Polish people suffer from asthma symptoms (Samoliński et al. 2014). Asthma is a heterogeneous disease. The guidelines emphasize the need to differentiate asthma phenotypes in terms of etiology and clinical issues (GINA 2017). Both asthma and COPD are complex conditions, characterized by a changeable clinical course (Guerra 2009; Kant and Gupta 2008; Seifart and Plagens 2007). They can be regarded as sharing many aspects of pathophysiology, clinical symptoms, and treatment (Gerhardsson et al. 2015). Shaya et al. (2008) have shown that therapy of these diseases requires a considerable financial outlay.

The incidence of respiratory diseases depends on genetic predisposition and environmental factors, such as allergens that favor the development of asthma, cigarette smoke, and the most important cause of COPD, atmospheric air pollution (Chen et al. 2005; Ronchetti et al. 2001; Kauffmann et al. 1996; El-Mehairy and El-Tarabishi 1966). Other contributors to the development of these conditions include socioeconomic factors, in particular poverty and malnutrition marked by a substantial weight loss (Kant and Gupta 2008). Apart from the environmental hazards, a global increase in the incidence of COPD and asthma can be underlain by genetic predisposition (Kalsheker and Chappell 2009; Wood and Stockley 2006). One of the potential genetic underling risk factors might be having a definite blood type. Hirszfeld and Hirszfeld (1919) demonstrated that the frequency of A and B blood types varies between populations. This observation has become a starting point for considering the relationship between the ABO blood groups and proneness to some diseases (Franchini and Lippi 2015; Franchini and Liumbruno 2013; Anstee 2010; Mourant et al. 1976). The ABO blood type antigens are complex carbohydrate molecules on the extracellular surface of the erythrocyte cell membrane. However, aside from their expression on erythrocytes, these antigens are also strongly expressed on the surface of various human cells and tissues, including the epithelium, sensory neurons, thrombocytes, and vascular endothelium. The clinical significance of the ABO blood groups thus goes beyond transfusion medicine. Several reports have suggested that this system plays a role in the development of cardiovascular, oncological, and other diseases (Cozi et al. 2017; Zhang et al. 2012).

The ABO blood type system may be a genetic factor associated with vulnerability to chronic respiratory diseases in some populations (Chen et al. 2005; Ronchetti et al. 2001; El-Mehairy and El-Tarabishi 1966; Kauffmann et al. 1996). The O and non-O blood types are regarded as genetic contributors to susceptibility to asthma. Denborough and Downing (1968) have sought for the confirmation of the relationship between ABO blood type agglutinins and the occurrence of respiratory allergic diseases, such as asthma and hay fever, in a group of 435 sick, compared with 411 healthy individuals. They show no appreciable differences in the secretory status or isoagglutinin titer in the serum and saliva depending on the ABO blood types. A similar study, carried out by Brachtel et al. (1979) has compared the blood types of children and adults with asthma and found that A and B blood types are related to various atopic states. Baiardini et al. (2008) have demonstrated that asthma and allergic rhinitis share analogous immunopathologic mechanisms, which indicates that both diseases are manifestations of a single syndrome with a wide spectrum of severity. Bijanzadeh et al. (2011) and Togias (2003) have suggested that genetic and environmental determinants of asthma also entail greater vulnerability to allergic rhinitis.

The majority of studies that consider the potential influence of blood typing on the risk of chronic respiratory diseases date from before 2000, and the most recent reports focus on childhood asthma, which, according to WHO (2003), is one of the most common chronic childhood diseases. Denborough and Downing (1968) have described a significant difference in the frequency of ABO blood types and related secretory phenotypes in adults with asthma. Similar findings have been reported for COPD by Cohen et al. (1977) who have indicated that the absence of blood type B and the presence of type A are related to COPD in white individuals. Other studies have demonstrated a higher frequency of A and B erythrocyte phenotypes in adult patients with various atopic states than that in the control groups. Additionally, O blood type has been found to entail proneness to asthma among Taiwanese and Italian children (Chen et al. 2005; Ronchetti et al. 2001), as well as adult Europeans (Kauffmann et al. 1996). Evidence of a connection between blood types and susceptibility to COPD and asthma is not sufficient to draw explicit conclusions (Liumbruno and Franchini 2013). Therefore, in this study we seek to determine the relationship between the ABO blood groups and the incidences of asthma and COPD in adult patients. Our working hypothesis in the study was that asthma and COPD could be associated with a definite blood type.

2 Methods

The research was performed in line with the Declaration of Helsinki and was approved by the Bioethics Committee of the Pomeranian Medical University in Szczecin, Poland (approval no. KB-0012/20/16).

2.1 Study Design

This study was a retrospective review of the statistical data from the Asseco Medical Management Solutions (AMMS), a public institution that runs an integrated and comprehensive system for hospitals and clinics in Poland, and from medical files of patients with the diagnosis of COPD or asthma, including those suffering from a cancerous disease with concomitant asthma or COPD, hospitalized from January 2016 to July 2017. The analysis concerned the patients whose digital and hardcopy files included information on the ABO blood groups. Data were gathered from 248 patients. The patients were hospitalized in the tuberculosis and pulmonary disease ward (95 patients; 38.3%), general and transplantation surgery ward (49 patients; 19.8%), and urology and urological oncology ward (104 patients; 41.9%) in the Public Provincial Hospital in the city of Szczecin, Poland. The inclusion criterion was the diagnosis of at least one chronic respiratory disease (COPD or asthma) as the main disease, or as a condition coexisting with cancerous diseases. Patients' sex, age, blood type, and comorbidities were obtained from a review of medical files.

2.2 Sociodemographic and Health Structure of Study Sample

Table 1 illustrates the structure of the study sample, including the diagnosis and blood types. The median age of patients was 65 years (min–max: 20–95 years), with no appreciable difference

Table 1 Sociodemographic and clinical features of study patients ($n = 248$)

Age (years)	
Median	65
Q_1–Q_3	60–70
Min–max	20–95
n (%)	
Gender	
Female	85 (34.3)
Male	163 (65.7)
Blood type	
A	97 (39.1)
B	56 (22.6)
O	69 (27.8)
AB	26 (10.5)
COPD	177 (71.4)
Asthma	71 (28.6)
Cancerous diseases	
Renal cancer	42 (34.7)
Bladder cancer	7 (5.8)
Prostate cancer	22 (18.2)
Gastric cancer	3 (2.5)
Pancreatic cancer	3 (2.5)
Colon cancer	18 (14.8)
Breast cancer	26 (21.5)

between men and women. The majority of patients (177; 67.0%) suffered from COPD and asthma (71; 33.0%). One hundred and twenty one patients were hospitalized for cancerous diseases, of whom 40 (33.1%) had asthma and 81 (66.9%) had COPD.

2.3 Statistical Analysis

Data are shown as medians with upper and lower quartiles and minimum-maximum ranges. Distribution of continuous variables was determined using the Shapiro–Wilk normality test. Categorical data are shown as the number and percentage. The Wilcoxon rank–sum test, chi–square test, and Fisher exact test were employed to compare proportions. The odds ratio (OR) with 95% confidence intervals were calculated. The statistical significance was set to alpha = 5% ($p < 0.05$). The analysis was performed using the statistical package R version 3.4.0.

3 Results

The A, B, O, and AB blood types were found in 39.1%, 22.7%, 27.8%, and 10.5% of the patients, respectively. These results differed from those obtained for the general Polish population, where blood type B is found in 17% and O in 37%. The differences in the distribution of blood types between the study sample and the general population were significant ($p = 0.007$).

Asthma was more common among younger patients (below 65 years of age) and COPD was among those over 65 years of age. Over two thirds (49; 69.0%) of asthma patients and less than half of COPD patients (79; 44.6%) aged ≤65. The difference in the age-related prevalence of both diseases was significant ($p < 0.001$) (Table 2).

The distribution of individual blood types in the asthma group did not differ from that in the COPD group of patients ($p = 0.717$) (Table 3). The blood types that predominated in asthma were A (26; 36.6%) and O (22; 31.0%) ($p = 0.02$) and in COPD they were A (71; 40.1%), O (47; 26.6%), and B (42; 23.7%) ($p < 0.001$). We then divided the blood types into two general categories: O and non-O (including A, B, and AB). The non-O category predominated in both asthma ($Chi^2 = 10.27$, $p = 0.001$) and COPD groups ($Chi^2 = 38.92$, $p < 0.001$). Hence, we draw a conclusion that the blood type is not a prognostic differentiating between the occurrence of asthma or COPD, although diseases are less common in people with blood type O.

Cancerous disease was observed in 121 (48.8%) of patients. Forty (56.3%) patients with asthma and 81 (45.8%) with COPD had a concomitant cancerous disease. There was no relationship between the incidence of asthma and cancerous disease ($p = 0.286$) or between

Table 2 Relationship between age and sex, and the incidence of asthma, COPD, and cancerous diseases

	Age ≤ 65; n (%)	Age > 65; n (%)	*p <	OR	95% CI
Asthma	49 (69.0)	22 (31.0)	0.001	0.36	0.19–0.67
COPD	79 (44.6)	98 (55.4)	0.001	2.75	1.49–5.21
Cancerous diseases	64 (52.9)	57 (47.1)	0.705	0.91	0.53–1.54
Gender					
Female	45 (52.9)	40 (47.1)	0.302	1.37	0.74–2.51
Male	83 (50.9)	80 (49.1)			

*Fisher's exact test; *OR* odds ratio; 95% *CI* 95% confidence intervals

Table 3 Blood type frequency in asthma and chronic obstructive pulmonary disease (COPD)

	Asthma	COPD	Total
Blood type	n (%)	n (%)	n (%)
A	26 (26.8)	71 (73.2)	97 (38)
B	14 (25.0)	42 (75.0)	56 (17)
O	22 (31.9)	47 (68.1)	69 (38)
AB	9 (34.6)	17 (65.4)	26 (8)
Total	71	177	248

Fisher's exact test: $p = 0.717$; $Chi^2 = 9.96$, $df = 3$, $p = 0.02$ (asthma); $Chi^2 = 33.24$, $df = 3$, $p < 0.001$ (COPD)

COPD and cancerous disease ($p = 0.160$). However, we found a significant difference in the incidence of cancerous disease depending on the blood type. Cancerous disease was more frequent among patients with the non-O (A and B) blood types (86; 71.1% patients) than in the O group (35; 28.9% patients) ($p < 0.001$).

3.1 Odds Ratios

The odds ratio for COPD in patients with non-O blood type was significantly higher than in those with blood type O (OR = 1.24, 95% CI: 1.64–2.36), with the 73.4% and 26.6% of patients with either blood type, respectively ($p < 0.05$). The odds ratio for asthma was about the same in patents with non-O and O blood types (OR = 0. 81, 95% CI: 0.42–1.56), with the 27.4% and 31.9% of patients with either blood type, respectively ($p < 0.05$).

The odds ratio for asthma in younger patients (≤ 65 years) was significantly higher than in those ≥65 years of age (OR = 1.36, 95% CI: 0.19–1.67; $p < 0.001$), with the 38.3% and 18.3% of patients of either age category, respectively ($p < 0.001$). The odds ratio for COPD in patients ≥65 years of age was nearly three times as higher as that in younger patients (≤ 65 years) (OR = 2.75, 95% CI: 1.49–5.21), with the 81.7% and 61.7% of patients of either age category, respectively ($p < 0.001$).

The odds ratio for COPD was the same in women and men (OR = 1.37, 95% CI: 0.74–2.51), with the 67.1% and 73.6% of female and male patients, respectively ($p = 0.302$). Likewise, the odds ratio for asthma was the same in women and men (OR = 0.73, 95% CI: 0.40–1.35), with the 32.9% and 26.38% % of female and male patients, respectively ($p = 0.302$). The odds ratio for cancerous disease in women was about three times as high as that in men (OR = 2.90, 95% CI: 1.63–5.24) ($p < 0.001$) and it was similar in both age-groups (OR = 0.91, 95% CI: 0.53–1.54).

4 Discussion

The working hypothesis in this study was that respiratory diseases could associate with ABO blood groups. However, the findings failed to support this hypothesis. The only significant finding was that pooled non-O blood types, i.e., A, B, and AB, predominated in patients with asthma as well as COPD. The blood group system is a genetically determined characteristic of the immune system. The system plays an essential, widely described role in transplantology and hemotherapy, but the role it could play in the development of other diseases, particularly of the inflammatory, infectious, or neoplastic vein, is not entirely understood (Guerra 2009; Garratty 2005). The present study was concerned with the plausible relation of two chronic respiratory disorders, COPD and asthma, with the blood groups. Particularly, COPD is a condition that has barely been studied in this context (Cohen et al. 1980; Kauffmann et al. 1996). Blood type B and O distribution in our study sample differed from the normal distribution assumed for Poland (22% *vs*. 17% for B, and 27.8% *vs*. 37% for O). Most asthma patients had blood type A or O, the differences between the two was significant to the advantage of A ($p = 0.02$), while the blood types in COPD patients were more diverse, including A, O, and B ($p < 0.001$). We also found that COPD and asthma were more common among patients with the non-O blood types (i.e., A, B, and AB).

A study conducted by Chen et al. (2005) in Taiwan, encompassing 136 asthmatic children and 161 control subjects, has revealed that blood type O is conducive to the occurrence of asthma. The authors conclude that O may contribute to the development of environmental allergies in children. In contradistinction, Bijanzadeh et al. (2009) have reported that blood groups do not appreciably associates with asthma in the populations of Mysore, Karnataka, and South India. Liumbruno and Franchini (2013) have reviewed the subject and conclude that there is evidence that the blood group system may be at play in various cardiovascular, neoplastic, and infectious disorders. However, diverse and sometimes opposing findings make it hard to draw a definite judgment. On the association side, Wu et al. (2008) have found that non-O blood types are related to vascular disorders, with OR of 1.25, which was confirmed by Franchini et al. (2013) who have described an increased risk of ischemic heart disease in the carriers of these blood types. The incidence of cancer and the ABO blood types has attracted understandable attention. Joh et al. (2012) have reported a non-significant association between non-O blood type and increased risk of renal cell cancer in a combined group of men and women; the association was significant in women evaluated separately. Suadicani et al. (2007) have carried out a 16-year-long study to assess the risk of death due to lung cancer, lifestyle, and occupational factors with respect to the ABO phenotype in a group of 3346 men aged 53–74 years. The authors have found that 170 (5.1%) of men died from lung cancer and the death rate was significantly greater in men with blood type O (84; 5.9%) than those with type A (70; 4.9%, or types B/AB (16; 3.2%). In contradistinction, Jaff (2010) has indicated that individuals with blood type O suffer from lung cancer less often than those with type A, which suggests that having type O might provide some protection against lung cancer.

Genetic and epidemiological studies of COPD patients focusing on such variables as age, sex, race, and smoking have pointed to the multifactorial background of the disease (Seifart and Plagens 2007). The presence of blood type A or the absence of type B are quoted by some authors as possible risk factors (Cohen et al. 1980). The knowledge that a multitude of genetic and environmental factors may influence the intensity of pulmonary disorders makes it important to seek for more specific phenotypes that can be tied to the ABO genotype; as exemplified by worse outcomes in females in case of renal cell carcinoma (Joh et al. 2012).

In the present study, we found no statistically significant gender-dependent differences in the distribution of the ABO blood types between men and women. Kauffmann et al. (1996) have assessed the influence of ABO genotype on respiratory variables in patients with chronic lung diseases and demonstrated that the lungs of patients with blood type A function worse. Ronchetti et al. (2001) have investigated the ABO/Secretor phenotype in asthmatic children, which controls mucosal secretion and adhesion of infectious agents in bronchial epithelium, and found that a proportion of O/nonSecretor was greater than that in healthy subjects. The association between O/nonSecretor phenotype and asthma has also been found in adult asthmatics (Kauffmann et al. 1996). Chen et al. (2005) have indicated that blood type O combined with the Sese gene is a risk factor for asthma in children (OR = 2.70; 95% CI: 1.13–6.03). Ronchetti et al. (2001), on the other hand, have demonstrated a positive correlation between the possibility of developing asthma in children with non-O blood types and the Sese gene during the first 3 years of life. El-Mehairy and El-Tarabishi (1966) have assessed the influence of blood types on the incidence of allergic diseases, including bronchial asthma, in a study of 500 patients. They noticed that blood type AB is more common among the unhealthy participants (7.45%) than in healthy controls (14.2%). In the present study, the majority of individuals with asthma and COPD had blood type A or O. However, non-O blood type visibly predominated over type O among patients with asthma and COPD. The differences between our results and the findings reported by El-Mehairy and El-Tarabishi (1966) may be caused by the difference in blood type distribution between races. Falsarella et al. (2011) have evaluated gender differences concerning the blood types and susceptibility to allergenic factors in adults. The authors reported a significant difference in the frequency of blood type phenotypes between males with allergic rhinitis and healthy males. No such relationship is observed in females. In males, phenotype A protects against allergic rhinitis (OR = 0.44, 95% CI: 0.20–0.94; $p = 0.049$) and blood type O predisposes to it (OR = 2.79, 95% CI: 1.38–5.62; $p = 0.01$).

Our results show that men suffered from COPD more often than women ($p = 0.048$). Another determinant of asthma and COPD was patient age. Asthma was more common among younger people (< 65 years of age) and COPD among those over 65 years of age ($p < 0.001$). Likewise, Rodriguez-Pecci et al. (2012) have found that men are more often former or current smokers, suffer several-fold more often from COPD, and if they do, they are in severer stages of the disease than women. However, women have a greater prevalence for COPD without exposure to cigarette smoke than men have (64.7 *vs.* 7.2%, respectively; $p < 0.001$), which may suggest that women are more sensitive to the development of the disease. That notion finds support in a study of Silverman et al. (2001) who show that women exposed to cigarette smoke in childhood are more likely to suffer from COPD.

We conclude that non-O blood types, i.e., A, B, and AB, predominate in chronic obstructive respiratory diseases, such asthma and COPD. The corollary is that both asthma and COPD are less common among people with blood type O. However, the blood type is not a prognostic differentiating between the occurrence of asthma or COPD. Further exploration is required concerning the predisposing role of the ABO antigens for chronic respiratory diseases. Nonetheless, unraveling the possible relationship between the blood group and the pathogenesis of respiratory diseases would enhance the predictive power for the development of such diseases.

Conflicts of Interest The authors declare no conflicts of interest in relation to this article.

References

Anstee DJ (2010) The relationship between blood types and disease. Blood 115(23):4635–4643

Baiardini I, Braido F, Tarantini F, Porcu A, Bonini S, Bousquet PJ, Zuberbier T, Demoly P, Canonica GW (2008) ARIA-suggested drugs for allergic rhinitis: what impact on quality of life? A GA2LEN review. Allergy 63(6):660–669

Bijanzadeh M, Ramachandra NB, Mahesh PA, Savitha MR, Manjunath BS, Jayaraj BS (2009) Lack of association between asthma and ABO blood type. Lung 187 (6):389–392

Bijanzadeh M, Mahesh PA, Ramachandra NB (2011) An understanding of the genetic basis of asthma. Indian J Med Res 134(2):149–161

Brachtel R, Walter H, Beck W, Hilling M (1979) Associations between a topic diseases and the polymorphic systems ABO, Kidd, Inv and red cell acid phosphatase. Hum Genet 49(3):337–348

Castro-Rodriguez JA, Holberg CJ, Wright AL, Martinez FD FD (2000) A clinical index to define risk of asthma in young children with recurrent wheezing. Am J Respir Crit Care Med 162:1403–1406

Chen Y, Chen JC, Lin TM, Huang TJ, Wang ST, Lee MF, Wang JY (2005) ABO/secretor genetic complex is associated with the susceptibility of childhood asthma in Taiwan. Clin Exp Allergy 35(7):926–932

Cohen BH, Ball WC, Jr BS, Diamond EL, Kreiss P, Levy DA, Menkes HA, Permutt S, Tockman MS (1977) Risk factors in chronic obstructive pulmonary disease (COPD). Am J Epidemiol 105(3):223–232

Cohen BH, Bias WB, Chase GA, Diamond EL, Graves CG, Levy DA, Menkes H, Meyer MB, Permutt S, Tockman MS (1980) Is ABH nonsecretor status a risk factor for obstructive lung disease? Am J Epidemiol 111:285–291

Cozzi GD, Levinson RT, Toole H, Snyder MR, Deng A, Crispens MA, Khabele D, Beeghly-Fadiel A (2017) Blood type, ABO genetic variants, and ovarian cancer survival. PLoS ONE 12(4):e0175119

Denborought MA, Downing HJ (1968) Secretor status in asthma and hay fever. J Med Genet 5(4):302–305

El-Mehairy MM, El-Tarabishi N (1966) The correlation between ABO blood types and allergic disorders. Ann Allergy 1966(24):366–368

Falsarella N, da Costa Ferreira AI, Nakashima F, de Brandao C, de Mattos C, de Mattos LC (2011) Evidence of an association between the O blood type and allergic rhinitis. Rev Bras Hematol Hemoter 33 (6):444–448

Franchini M, Lippi G (2015) The intriguing relationship between the ABO blood type, cardiovascular disease, and cancer. BMC Med 13:7

Franchini M, Liumbruno GM (2013) ABO blood type: old dogma, new perspectives. Clin Chem Lab Med 51:1545–1553

Franchini M, Rossi C, Mengoli C, Frattini F, Crestani S, Giacomini I, Luppi M, Bonfanti C (2013) ABO blood type and risk of coronary artery disease. J Thromb Thrombolysis 36:286–287

Garratty G (2005) Relationship of blood types to diseases: do blood type antigens have a biological role? Rev Med Inst Mex Seguro Soc 43(1):113–121

Gerhardsson de Verdier M, Andersson M, Kern DM, Zhou S, Tunceli O (2015) Asthma and chronic obstructive pulmonary disease overlap syndrome: doubled costs compared with patients with asthma alone. Value Health 18(6):759–766

GINA (2017) Global initiative for Asthma. Global strategy for Asthma management and prevention. http://ginasthma.org/2017-gina-report-global-strategy-for-asthma-management-and-prevention/. Accessed 10 Aug 2017

Global Disease Burden (2013) http://global-disease-burden.healthgrove.com/l/46197/Chronic-Obstructive-Pulmonary-Disease-in-Poland. Accessed on 08 Mar 2018

Guerra S (2009) Asthma and chronic obstructive pulmonary disease. Curr Opin Alergy Clin Immunol 9 (5):409–416

Hirszfeld L, Hirszfeld H (1919) Serological differences between the blood of different races. Lancet 197:675–679

Jaff MS (2010) Higher frequency of secretor phenotype in O blood type – its benefits in prevention and/or treatment of some diseases. Int J Nanomedicine 5:901–905

Joh HK, Cho E, Choueiri TK (2012) ABO blood type and risk of renal cell cancer. Cancer Epidemiol 36:528–532

Kalsheker N, Chappell S (2009) The new genetics and chronic obstructive pulmonary disease. J COPD 5:257–264

Kant S, Gupta B (2008) Role of lifestyle in the development of chronic obstructive pulmonary disease: a review. Lung India 25(2):95–101

Kauffmann F, Frette C, Pham QT, Nafissi S, Bertrand JP, Oriol R (1996) Associations of blood type-related antigens to FEV1, wheezing, and asthma. Am J Respir Crit Care Med 153(1):76–82

Kim SR, Rhee YK (2010) Overlap between asthma and COPD: where the two diseases converge. Allergy Asthma Immunol Res 2(4):209–214

Liumbruno G, Franchini M (2013) Beyond immunohaematology: the role of the AB0 blood type in human diseases. Blood Transfus 11:491–499

Mathers CD, Loncar D (2006) Projections of global mortality and burden of disease from 2002 to 2030. PLoS Med 3(11):209–224

Mourant AE, Kopec AC, Domaniewska-Sobczak K (1976) The distribution of the human blood groups and other polymorphisms, 2nd edn. Oxford University Press, London

Rodriguez-Pecci MS, de la Fuente-Aguado J, Montero-Tinnirello, Sanjurjo-Rino AB, Sánchez-Conde P, Fernández-Fernández FJ (2012) Chronic obstructive pulmonary disease: differences between men and women. Medicine (B Aries) 3:207–215 (Article in Spanish)

Ronchetti F, Villa MP, Ronchetti R, Bonci E, Latini L, Pascone R, Bottini N, Gloria-Bottini F (2001) ABO/secretor genetic complex and susceptibility to asthma in childhood. Eur Respir J 17(6):1236–1238

Samoliński B, Raciborski F, Lipiec A, Tomaszewska A, Nowicka A (2014) Epidemiology of allergic diseases in Poland. Pol J Allergol 1:10–18

Seifart C, Plagens A (2007) Genetics of chronic obstructive pulmonary disease. Int J Chron Obstruct Dis 2 (4):541–550

Shaya FT, Dongyi D, Akazawa MO, Blanchette CM, Wang J, Mapel DW, Dalal A, Scharf SM (2008) Burden of concomitant asthma and COPD in a Medicaid population. Chest 134(1):14–19

Silverman EK, Weiss ST, Drazen JM, Chapman HA, Carey V, Campbell EJ, Denish P, Silverman RA, Celedon JC, Reilly JJ, Ginns LC, Speizer FE (2001) Gender-related differences in severe, early-onset chronic obstructive pulmonary disease. Am J Respir Crit Care Med 162(6):2152–2158

Suadicani P, Hein HO, Gyntelberg F (2007) ABO phenotypes and inflammation-related predictors of lung cancer mortality: the Copenhagen male study – a 16-year follow-up. Eur Respir J 30:13–20

Togias A (2003) Rhinitis and asthma: evidence for respiratory system integration. J Allergy Clin Immunol 111 (6):1171–1183

WHO (2003) Management of non communicable diseases department: prevention of Allergy and Allergic Asthma WHO; http://apps.who.int/iris/bitstream/10665/68361/1/WHO_NMH_MNC_CRA_03.2.pdf. Accessed on 8 Aug 2017

WHO (2017) Chronic respiratory diseases. http://www.who.int/gard/publications/chronic_respiratory_diseases.pdf. Accessed on 8 Aug 2017

Wood AM, Stockley RA (2006) The genetics of chronic obstructive pulmonary disease. Respir Res 7:130

Wu O, Bayoumi N, Vickers MA, Clark P (2008) ABO (H) blood types and vascular disease: a systematic review and meta-analysis. J Thromb Haemost 6:62–69

Zhang H, Mooney CJ, Reilly MP (2012) ABO blood groups and cardiovascular diseases. Int J Vasc Med 2012:641917

Advs Exp. Medicine, Biology - Neuroscience and Respiration (2018) 41: 41–47
https://doi.org/10.1007/5584_2018_242

Published online: 27 July 2018

Satisfaction with Life and Adaptive Reactions in People Treated for Chronic Obstructive Pulmonary Disease

Bartłomiej Drop, Mariola Janiszewska, Agnieszka Barańska, Krzysztof Kanecki, Aneta Nitsch-Osuch, and Magdalena Bogdan

Abstract

Chronic obstructive pulmonary disease (COPD) is a common illness. An individualized approach to changes caused by a chronic disease includes specific activity resulting from cognitive-emotional predispositions. The purpose of this study was to determine the influence of selected factors such as satisfaction with life, health self-assessment, duration of the illness, and the way of coping with disease on adaptive reactions in COPD patients. The study involved 109 COPD patients. The *Adaptive Responses* Inventory and the *Satisfaction with Life Scale* were used for the assessment of variables. We found that the majority of patients had low or very low satisfaction with life (63.3% and 22.0%, respectively). There was an inverse association between anxiety and satisfaction with life ($r = -0.279$; $p < 0.003$). Patients who assessed their health as good or very good showed a significantly greater satisfaction with life ($p < 0.001$). In the main, attitudes demonstrating a greater satisfaction with life and motivation to cope with COPD did not predominate among the patients. A poor state of mind, and stress inversely affected satisfaction. We conclude that although positive stress can mobilize to fight for better health, anxiety caused by illness has a negative connotation as it may demotivate to protect and maintain better health and consequently may reduce satisfaction with life.

Keywords

Adaptive reactions · COPD · Coping with disease · Health self-assessment · Satisfaction with life

B. Drop and A. Barańska
Department of Informatics and Medical Statistics, Medical University of Lublin, Lublin, Poland

M. Janiszewska
Chair of Public Health, Faculty of Health Science, Medical University of Lublin, Lublin, Poland

K. Kanecki, A. Nitsch-Osuch (✉), and M. Bogdan
Department of Social Medicine and Public Health, Medical University of Warsaw, Warsaw, Poland
e-mail: agnieszkazablocka2@wp.pl

1 Introduction

Chronic obstructive pulmonary disease (COPD) is a common illness which changes almost every sphere of the biopsychosocial functioning in patients. The disease makes the patient introduce radical changes in meeting daily needs, defining and selecting the ways to achieve goals, and shaping the competencies necessary for the optimal fulfilment of social roles. Disease-related limitations significantly change the way of life and become the basic context for defining the future. A dominant role is played by the awareness of disease consequences, current life situation, possessed abilities, remedial skills, and the help of loved ones. An individualized approach to changes caused by a chronic disease is related to cognitive-emotional predispositions. The adaptation to life with a chronic disease requires to determine a specific way of behaving in the face of objective changes caused by an illness, and their subjective evaluation in the context of a future life perspective. The relationship between satisfaction with life, as a measure of positive adaptation, and health, as a subjective assessment of one's own health could be considered the mediators of adaptive reactions. However, complexity and mutual dependence of the interaction between a patient and the environment, when an illness strikes, play an essential role in shaping satisfaction with life (Livneh 2001; Falvo and Holland 1999). A loss of health does not necessarily mean a reduced satisfaction with life (Nosek et al. 1995). COPD as a chronic disease is posed to affect the patient's life. Recent studies in COPD patients have shown that physical and functional limitations adversely affect satisfaction with life. There are, however, no reports on the importance of adaptive reactions to the disease in improving patients' quality of life. Therefore, this study investigates the hypothesis that reactions categorized as early (anxiety, denial, and shock) and intermediate (depression, hostility, and internal anger) would inversely associate with the satisfaction with life, health self-assessment, illness duration, and the way of coping with disease stress, and conversely, desirable adaptive reactions (acceptance and recognition) would positively associate with the psychosomatic features above outlined.

2 Methods

The study was approved by the Ethics Committee of the Medical University of Lublin, Poland (permission no. KE-0254/25/2016) and was conducted in accordance with the principles for human research set by the Declaration of Helsinki. One hundred and nine COPD patients (32.1% men and 67.9% women; mean age 63 ± 9, range 45–89 years) were enrolled after giving informed consent for the study procedures. The majority of patients (63.3%) were residents of cities. More than half (51.4%) were in a relationship (marriage or partnership), 8.3% were divorcees, 5.5% were singles, and 34.9% were widowed. Most of the patients (67.9%) assessed their social-living conditions as very good or good, while the remaining 32.2% assessed them as bad or very bad. The study was conducted in the Public Clinical Hospital No. 4 and in the outpatient clinic for lung diseases and tuberculosis of the *Jan Boży* Public Provincial Hospital in Lublin between February 2016 and July 2017.

The psychometric tools consisted of the *adaptive responses* inventory (Livneh and Antonak 2005) and the *satisfaction with life scale* (SWLS) (Diener et al. 1985) in Polish adaptations. The adaptive responses inventory enables the identification of the intensity of reactions demonstrated by patients as a result of limitations caused by a chronic illness. It consists of 8 subscales that express the following reactions: anxiety, shock, denial, internal anger, depression, manifested hostility, acceptance, and adaptation. The achievable score ranges from 1 to 5 points. The SWLS consists of five statements; the respondents report on a 7-point scale to what extent each statement relates to their current life. All the statements measure factors associated with satisfaction with life, and the score ranges from 5 to 35 points. The Cronbach alpha reliability indices of these tools are satisfactorily high; 0.78 and 0.81, respectively. The subjective assessment of a health condition, duration of illness, and the presence of secondary health complications was surveyed with an ancillary questionnaire developed for this research by the authors.

Continuous data were reported as means ±SD and the minimum–maximum range. These data were statistically compared with a *t*-test when normally distributed and with the Mann-Whitney

U or Kruskal-Wallis test when non-normally distributed. Categorical data were compared with Chi^2. Spearman's rho correlation coefficient was used for the evaluation of correlations. A p-value <0.05 defined statistical significance of differences. The analysis was performed using a commercial SPSS Statistics 20 package (IBM Corp., Armonk, NY).

3 Results

We found that over half of patients rated their health as bad (50.5%) or very bad (2.8%), while 45.9% rated it as good, and only 0.9% as very good. Among diseases coexisting with COPD, heart and cardiovascular disorders were the most frequent, including stroke and hypertension (38.5%), followed by gastroesophageal reflux disease (36.8%), pulmonary embolism (33.9%), diabetes mellitus and/or metabolic syndrome (33.9%), lung cancer (22%), and also anxiety and depression (9.2%). An important health issue in these patients is the co-existence of many extrapulmonary symptoms. The most notable complaints were exhaustion and wasting syndrome (44.0%), reduction in systemic and local immunity (31.2%), weakening of muscle strength (20.2%), and weight loss (20.2%).

Satisfaction with life is a cognitive component of subjective human well-being. It is the way in which individuals evaluate the quality of their life on the basis of their own unique set of criteria. As the disease generally causes negative effects in every sphere of life, it is justified to study satisfaction with life among chronically ill people. The SWLS reflects the long-term perspective rather than short-term, and therefore it concerns the global assessment of life rather than satisfaction with specific achievements or spheres of life (Pavot and Diener 1993). In the present study, the highest score in SWLS amounted to 22 out of the 35 points possible, none of the patients showed a high level of satisfaction with life. Most COPD patients declared low (63.3%) or very low (22.0%) satisfaction, and only 14.7% had the average level of it. Since most patients had a low level of satisfaction, it may be assumed that

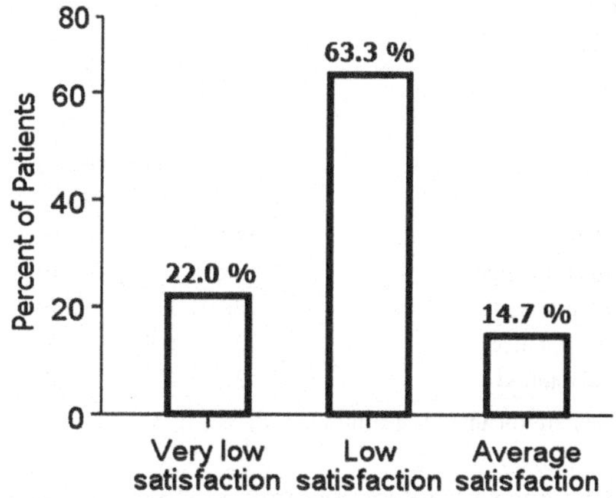

Fig. 1 Chronic obstructive disease (COPD) patients with different categories of satisfaction with life

they experienced a dissonance between expectations and their fulfilment (Fig. 1).

Concerning the adaptive reactions, patients with COPD most often responded to the disease by anger and hostility, manifested mainly toward the self, with a mean score of 3.5 ± 0.3 points, which likely was a consequence of faulting themselves for the disease appearing. The disease acceptance was fairly high among patients, the mean score of 3.2 ± 0.3 (Table 1). Thus, patients learned to live with the disease symptoms and show an impassive attitude to their predicament. The possible causes of this attitude might be that patients with advanced COPD often fail to cope with adherence to medical recommendations due to physical or intellectual disability, resulting from chronic hypoxia and depression or from social isolation, loneliness, and the feeling of abandonment. In some patients, the unfavorable course of COPD appears to be directly linked to financial shortfalls and low social status, and to the lack of support from loved ones. In such a case, patients generally seem to accept the annoying symptoms of the disease as "everyday humdrum", regarding an exacerbation as a "disease". Moreover, most patients cannot recognize early symptoms of exacerbations or react to them properly.

The lowest score in the *adaptive responses* inventory concerned the depression subscale (Table 1), which indicates that such non–adaptive

Table 1 Results of the satisfaction with life scale and the *adaptive responses* inventory

Satisfaction with life	14.0 ± 3.2 (8.0–22.0)
Shock	2.3 ± 0.2 (1.9–2.8)
Anxiety	2.6 ± 0.5 (1.6–3.6)
Denial	2.4 ± 0.2 (2.0–3.0)
Depression	1.6 ± 0.2 (1.3–2.0)
Internal anger	3.5 ± 0.3 (2.7–4.0)
Manifested hostility	2.9 ± 0.4 (1.8–3.6)
Acceptance	3.2 ± 0.3 (2.4–3.7)
Adaptation	3.0 ± 0.3 (2.3–3.8)

Data are means ±SD (min–max range)

Table 2 Correlations between the *satisfaction with life score* and adaptive reactions

	Satisfaction with life	
	Rho	p <
Shock	−0.269	0.001
Anxiety	−0.279	0.003
Denial	0.157	0.103
Depression	−0.233	0.004
Internal anger	−0.086	0.374
Manifested hostility	−0.238	0.004
Acceptance	−0.290	0.001
Adaptation	0.087	0.370

reactions were rather uncommon. However, some patients had evident depressive responses and the coexistence of depression and COPD should be taken into consideration. Depression, often undiagnosed, significantly decreases satisfaction with life and adversely affects the prognosis. At the same time, poor satisfaction with life is the strongest depression predictor (Antoniu 2011).

We found an inverse association between satisfaction with life and adaptive reactions. The stronger was the shock, anxiety, depression, and hostility toward others, the lesser the satisfaction with life. Likewise, a greater acceptance of disease consequences was associated with a smaller satisfaction with life. That demonstrates that the acceptance of a disease does not always indicate a positive adaptive reaction, as it may sometimes result from the lack of choice or help. The patient is forced to accept a situation which does not necessarily have a positive relationship with satisfaction with life (Table 2).

The subjective health assessment turned out a variable effectively differentiating satisfaction with life in patients. A significantly high satisfaction was shown by those who assessed their health as very good or good. The health self–assessment variable seems thus important in the evaluation of sometimes obscure nature of adaptive reactions. Patients with different health self–assessment reveal different adaptive reactions (Table 3). Low health self–assessment significantly more often was accompanied by denial ($p < 0.003$), depressive responses ($p < 0.001$), and hostility toward others ($p < 0.011$) than was high self–assessment, which, in turn, revealed a significantly higher acceptance of patient's limitations ($p < 0.014$) and adaptation to disease consequences ($p < 0.041$). Patients with low health self–assessment usually revealed reaction patterns, such as excessive anger and hostility, toward both themselves and others (Table 3). A high intensity of hostility toward others may result from the inadequate support the subjects received in relation to their needs and expectations. That most likely resulted from focusing on the disease cause and on the search for people to blame for disease consequences. These reactions significantly impede the acceptance of limitations, which is related to an increase in behavioral adaptation to life with COPD. The activation of aspirations to accept the current health abilities encounters a barrier in the form of the elevated level of depressive reactions.

It has been assumed that the duration of COPD and the sense of coping with its complications condition the relationships being analyzed. The assumption was also made that not just the presence of complications, but the subjective attitude toward them is essential for satisfaction with life and adaptive reactions. The duration of COPD treatment did not significantly differentiate the patients in terms of satisfaction with life and adaptive reactions (Table 4). However, satisfaction with life was somehow greater in patients who suffered from COPD for more than 5 years, compared with those who suffered for 1 or 2–5 years. The evaluation of adaptive reactions concerning the COPD duration showed no appreciable differences. However, patients who

Table 3 Satisfaction with life, adaptive reactions, and health self-assessment

	Good/very good	Bad/very bad	p <
Satisfaction with life	15.0 ± 3.3	13.0 ± 2.8	0.001
Shock	2.3 ± 0.2	2.3 ± 0.2	0.835
Anxiety	3.2 ± 0.3	3.3 ± 0.2	0.096
Denial	2.4 ± 0.5	2.7 ± 0.5	0.003
Depression	1.5 ± 0.2	1.7 ± 0.2	0.001
Internal anger	2.4 ± 0.2	2.4 ± 0.2	0.342
Manifested hostility	2.8 ± 0.4	3.0 ± 0.4	0.011
Acceptance	3.6 ± 0.3	3.4 ± 0.4	0.014
Adaptation	3.1 ± 0.3	3.0 ± 0.3	0.041

Data are means ±SD

Table 4 Satisfaction with life, adaptive reactions, and duration of illness

	Duration of COPD			
	1 year	2–5 years	> 5 years	p <
Satisfaction with life	13.7 ± 3.0	14.1 ± 3.7	14.3 ± 3.4	0.710
Shock	2.4 ± 0.2	2.4 ± 0.2	2.3 ± 0.2	0.132
Anxiety	2.7 ± 0.4	2.7 ± 0.5	2.5 ± 0.5	0.226
Denial	2.4 ± 0.2	2.4 ± 0.2	2.4 ± 0.2	0.833
Depression	1.6 ± 0.2	1.6 ± 0.2	1.6 ± 0.2	0.376
Internal anger	3.5 ± 0.3	3.5 ± 0.3	3.5 ± 0.3	0.415
Manifested hostility	3.0 ± 0.4	2.9 ± 0.5	2.9 ± 0.5	0.732
Acceptance	3.2 ± 0.3	3.2 ± 0.2	3.3 ± 0.2	0.549
Adaptation	3.0 ± 0.3	3.1 ± 0.4	3.0 ± 0.3	0.962

Data are means ±SD; *COPD* chronic obstructive pulmonary disease; Kruskal-Wallis test

suffered for a shorter time (up to 1 year) manifested greater shock, anxiety, denial, and even depression than those who suffered for more than 5 years. Patients who had time to learn about disease consequences were more likely to show acceptance and adaptation to the situation (Table 4).

The sense of coping with COPD complications significantly differentiated the patients in terms of satisfaction with life. Coping was considerably greater in patients who declared that they dealt with the complications well than in those who declared helplessness in this regard; 16.1 ± 2.9 *vs.* 13.8 ± 3.2 points ($p < 0.032$), respectively (Table 5). The former usually manifested the reactions of cognitive and emotional acceptance of limitations ($p = 0.001$) and behavioral adaptation to daily functioning. Nevertheless, in some situations these individuals tended to reveal excessive anger toward others or themselves. On the other hand, individuals who declared themselves helpless in confrontation with disease consequences reacted relatively often anxiety ($p < 0.001$), shock ($p < 0.037$), and even depressive responses ($p < 0.033$). It is worth noting that the lack of ability to cope with disease complications decreased the level of satisfaction with life (Table 5).

4 Discussion

The findings of this study disclosed various combinations of interactions between individual adaptive reactions and satisfaction with life in COPD patients. We found that both undesirable early adaptive reactions, such as shock and anxiety, and intermediate adaptive reactions, such as depression and manifested hostility, have an inverse relationship with satisfaction with life. An inverse nature of these relationships suggests that the responses, such as anger toward others

Table 5 Adaptive reactions and the sense of coping with the complications of COPD

	Yes	No	p <
Satisfaction with life	16.1 ± 2.9	13.8 ± 3.2	0.032
Shock	2.3 ± 0.2	2.5 ± 0.1	0.037
Anxiety	3.0 ± 0.3	3.3 ± 0.3	0.001
Denial	2.4 ± 0.1	2.4 ± 0.2	0.839
Depression	2.7 ± 0.5	3.0 ± 0.4	0.033
Internal anger	3.5 ± 0.3	3.4 ± 0.3	0.326
Manifested hostility	3.0 ± 0.4	2.7 ± 0.7	0.247
Acceptance	3.3 ± 0.2	2.9 ± 0.3	0.001
Adaptation	3.1 ± 0.4	3.0 ± 0.3	0.495

Data are means ±SD; *COPD* chronic obstructive pulmonary disease; Mann-Whitney U test

and depression, lower a positive mental attitude and thus satisfaction with life. In addition, the findings revealed that shock and anxiety also inversely associated with satisfaction with life. Although anxiety reactions, understood as the awareness of one's limitations, can mobilize the fight for health, uncontrolled and unpredictable anxiety caused by the illness can paralyze, reduce motivation for health protection, shape a subjective sense of disability, and consequently reduce satisfaction with life. Livneh et al. (2004) have also demonstrated that depression and anger are associated with the perception of mental health control.

On the other side, the adaptive reaction, expressed as willingness to accept disease symptoms, and treating the disease as one of the characteristics of one's personality, may not favor life satisfaction either. In COPD patients, this reaction revealed an inverse association with satisfaction. It is possible that the acceptance results from the lack of other possibilities and surrendering to the situation. Here, the acceptance is understood as an inability to solve the problem and the lack of strength to fight its consequences. This is a disturbing phenomenon, because it deepens health deterioration rather than encourages the improvement of one's condition. It also impacts the patient's mental condition. The literature indicates that this group of patients manifests up to several exacerbations requiring hospitalization each year (Lareau Yawn 2010; Breekveldt-Postma et al. 2007). Therefore, disease acceptance in many COPD patients is not necessarily a positive reaction. This association has not been demonstrated in other studies. However, a positive association has been reported between the acceptance of disability and the way of coping with it in spinal cord lesion patients (Elfström et al. 2005). The emerging predictive role of adaptive reactions in the variability of health-related practices may help control and manage the disease process (Suarez et al. 2013).

Further, the present study demonstrated a moderating role of disease duration and the way of coping with disease consequences, but also of health self-assessment in the adaptive reactions. Satisfaction with life tended to be better in patients who functioned with COPD for a longer time, even though the time that elapsed from onset of disease failed to clearly differentiate the patients' satisfaction. Changes in satisfaction with life in COPD patients, which occur over time, may intensify if they relate to the sphere of values and are associated with the reconstruction of the self-concept. The association of treatment duration and health self-assessment has been pointed out by Jensen et al. (2012). Krause et al. (2005) have also reported that the subjective health assessment is a significant factor in the adaptation process. A high level of health self-assessment, shorter duration of the illness, and the ability to cope with disease-related complications are variables important to the development of constructive adaptive reactions, i.e., emotional acceptance and adaptation. When the health self-assessment is poor, these reactions are expressed by denial, depression, and hostility toward others. These non-adaptive reactions are accompanied by poorer acceptance of disability restrictions (Duggan and Lysack 2001). The available evidence justifies making an assumption about the existence of such associations, although it is hard to explicitly specify their exact nature (Elfström et al. 2005; Livneh and Wilson 2003).

Recent literature reports argue that typical adaptive reactions, such as the acceptance of disease limitations associate with behaviors favoring physical and mental health, the quality of life, and personal satisfaction (Pollard and Kennedy 2007). Accordingly, non-adaptive reactions ought to show an inverse association. The present study confirmed these notions in COPD patients. However, both adaptive reactions and satisfaction with life have numerous intertwined associations, whose exact nature is still conjectural, particularly in chronically ill people. It is unknown, for instance, how a greater openness of an ill person to others, and *vice versa* others to the ill person, influences satisfaction with life. As this factor remains in the sphere of subjective human feelings, there is no possibility of its direct observation. Further, the associations of adaptive responses and satisfaction with life are dynamically structured as their background is individually shaped (Drum 2014). We should acknowledge that the level of satisfaction depends on multiple factors and is different in each person.

We conclude that attitudes demonstrating a greater life satisfaction and motivation to cope with COPD did not predominate among the patients. Although positive stress can mobilize to fight for better health, anxiety caused by illness has a negative connotation as it may demotivate to protect and maintain health and, in consequence, decreases life satisfaction. On the other side, it is worthwhile to note that willingness to accept disease symptoms may also inversely associate with satisfaction with life. The acceptance may reflect psychological surrender to the pathological condition, which actually is a further health detriment.

Conflicts of Interest The authors declare no conflicts of interest in relation to this article.

References

Antoniu SA (2011) Predictors of depression in chronic obstructive pulmonary disease patients. Expert Rev Respir Med 5:333–335

Breekveldt-Postma NS, Koerselman J, Erkens JA, Lammers JW, Herings RM (2007) Enhance persistence of tiotropium compared with other respiratory drugs in COPD. Respir Med 101:1398–1405

Diener E, Emmons RA, Larsen RJ, Griffin S (1985) The satisfaction with life scale. J Pers Assess 49:71–75

Drum CE (2014) The dynamics of disability and chronic conditions. Disabil Health J 7(1):2–5

Duggan CH, Lysack C (2001) How individuals interpret changes in quality of life following traumatic spinal cord injury. J Spinal Cord Med 24(Suppl 1):34

Elfström ML, Ryden A, Kreuter M, Taft C, Sullivan M (2005) Relations between coping strategies and health-related quality of life in patients with spinal cord lesion. J Rehabil Med 37(1):9–16

Falvo DR, Holland E (1999) Medical and psychosocial aspects of chronic illness and disability. Jones & Bartlett Learning, Burlington

Jensen MP, Molton IR, Groah SL, Campbell ML, Charlifue S, Chiodo A, Forchheimer M, Krause JS, Tate D (2012) Secondary health conditions in individuals aging with SCI: terminology, concepts and analytic approaches. Spinal Cord 50(5):373–378

Krause JS, Stanwyck CA, Maides J (2005) Locus of control and life adjustment. Relationship among people with spinal cord injury. Rehabil Couns Bull 41 (3):162–172

Lareau SC, Yawn BP (2010) Improving adherence with inhaler therapy in COPD. Int J Chron Obstruct Pulmon Dis 5:401–406

Livneh H (2001) Psychosocial adaptation to chronic illness and disability: a conceptual framework. Rehabil Couns Bull 44(1):151–160

Livneh H, Antonak RF (2005) Psychosocial adaptation to chronic illness and disability: a primer for counselors. J Couns Dev 83:12–20

Livneh H, Wilson LM (2003) Coping strategies as predictors and mediators of disability-related variables and psychosocial adaptation: an exploratory investigation. Rehabil Couns Bull 46(1):194–208

Livneh H, Lott SM, Antonak RF (2004) Patterns of psychosocial adaptation to chronic illness and disability: a cluster analytic approach. Psychol Health Med 9 (4):411–430

Nosek M, Furher M, Potter C (1995) Life satisfaction of people with physical disabilities: relationship to personal assistance, disability status and handicap. Rehabil Psychol 40(3):191–202

Pavot W, Diener E (1993) Review of the satisfaction with life scale. Psychol Assess 5:164–172

Pollard C, Kennedy P (2007) A longitudinal analysis of emotional impact, coping strategies and post-traumatic psychological growth following spinal cord injury: a 10-year review. Br J Health Psychol 12(Pt 3):347–362

Suarez NC, Levi R, Bullington J (2013) Regaining health and wellbeing after traumatic spinal cord injury. J Rehabil Med 45(10):1023–1027

Advs Exp. Medicine, Biology - Neuroscience and Respiration (2018) 41: 49–55
https://doi.org/10.1007/5584_2018_201

Published online: 21 April 2018

Estimates of Medication Expenditure for Ischemic Heart Disease Accompanying Chronic Obstructive Pulmonary Disease

Edyta Rysiak, Izabela Prokop, Ilona Zaręba, and Robert M. Mróz

Abstract

Ischemic heart disease (IHD) is a frequent accompaniment of chronic obstructive pulmonary disease (COPD). Co-occurrence of these two diseases is associated with many risk factors, difficulties in implementing appropriate therapies, numerous complications, and high spending for treatment. All these elements significantly reduce the quality of life of patients. The aim of this study was to estimate the expenditure for medications involved with IHD pharmacotherapy in the course of COPD. This retrospective study was based on the review of medical files of 57 patients, 27 women and 30 men, diagnosed with IHD, according to the severity classification, in the course of COPD which was staged according to the GOLD criteria. We found a considerable increase in *per capita per year* retail spending for drugs. The spending increased with the severity class of IHD; from 27.41 EUR in Class I to 142.30 EUR in Class IV. This spending did not include the treatment cost for the basic disease, i.e., COPD. A high individual cost burden was decreased by a discounting intervention of the National Health Fund. Despite a relatively high drug expenditure, we consider the treatment being cost-effective since we noticed a reduction in the classical risk factors for IHD, related to metabolic disturbances and lifestyle features, as soon as 2 months after treatment initiation. This study confirms that heart disease accompanying COPD is a frequent occurrence, generating high costs of treatment, which relates to the severity of this comorbidity.

Keywords

COPD · Ischemic heart disease · Medical costs · Pharmacokinetics

1 Introduction

Chronic obstructive pulmonary disease (COPD), one of a major global health problems, is characterized by an incompletely reversible airflow limitation which is usually progressive and is associated with an abnormal pulmonary inflammatory response to harmful particles or gases. In addition to pulmonary manifestations, i.e., progressive dyspnea and chronic cough COPD is a systemic disease that manifests itself in increasing

E. Rysiak (✉), I. Prokop, and I. Zaręba
Department of Medicinal Chemistry, Faculty of Pharmacy with the Division of Laboratory Medicine, Medical University of Bialystok, Białystok, Poland
e-mail: edyta.rysiak@umb.edu.pl

R. M. Mróz
Second Department of Lung Diseases and Tuberculosis, Medical University of Bialystok, Białystok, Poland

Respiratory Medicine Center, Białystok, Poland

weakness, cachexia, and the development of comorbidities. Therefore, appropriate therapy increases patients' quality of life and optimizes healthcare costs (GOLD 2016; Chatila et al. 2008; Fabbri et al. 2008; Gudmundsson et al. 2006; van Manen et al. 2001).

Cardiovascular diseases are one of the most common cause of death worldwide, ranking first on the list of causes of death in Europe. According to the Polish Cardiac Society, nearly 500 people die each day from cardiovascular diseases in Poland, of which 20% are under 65 years of age (Zdrojewski et al. 2013). Ischemic heart disease (IHD) considerably limits patient's daily activities and quality of life.

A risk of developing IHD in the course of COPD increases, particularly with the extending COPD duration. The prognosis of COPD combined with IHD is poorer than those for either disease considered separately (GOLD 2016; Curkendall et al. 2006; Soriano et al. 2005; Fear et al. 2010; Buist et al. 2007; Mannino et al. 2008; Yusuf et al. 2004). Both diseases have a mutually potentiating effect on each other, resulting in enhanced patient disability, with numerous undesirable health effects (Campo et al. 2013; Enriquez et al. 2011; Bursi et al. 2010; Sin and Man 2005).

Despite the relationship between COPD and IHD, there are no common standards in the mutual definition of these disease (Soriano et al. 2010; Brekke et al. 2008). The increasing number of patients simultaneously suffering from COPD and IHD in productive age is not only a medical issue but also a significant social and economic burden. The aim of this study was to evaluate the effectiveness and cost of pharmacotherapy of ischemic heart disease in patients with COPD.

2 Methods

This is a retrospective study based on a review of medical files of 57 patients (30 men and 27 women) diagnosed with, and treated for, ischaemic heart disease and COPD in the Respiratory Medicine Center, Bialystok, Poland between June and December 2016. Due to the retrospective nature of the anonymized patients' files, the institutional review board waived the requirement of obtaining ethical permission to conduct the study. The survey included the files of the patients diagnosed and treated according to the GOLD criteria for COPD. A history of cardiovascular diseases, COPD, family history, and lifestyle information were collected from all patients. The biochemical tests considered consisted of fasting lipidogram and glucose level in the serum. Also, anthropometric measurements, arterial pressure, and laboratory tests were gathered from medical records.

The following variables were evaluated:

- degree of severity of COPD;
- BMI (kg/m^2);
  - normal body mass: BMI <25 kg/m^2,
  - overweight: BMI = 25–30 kg/m^2,
  - obesity: BMI > 30 kg/m^2;
- physical activity: exercise at least 3–4 times a week for at least 30–40 min each time;
- alcohol use – consumption of 1–2 alcohol units or more at least twice a week (1 unit = 250 ml of bear, 150 ml of wine, or 30 ml of hard liquor);
- tobacco smoking: current and former;
- history of diabetes: glycemia over 120 mg% in duplicate examinations;
- history of arterial hypertension: previous diagnosis or arterial pressure $\geq$ 140/90 mmHg at two repeat examinations;
- history of hyperlipidemia, according to the following criteria: total cholesterol >200 mg/dl, LDL >115 mg/dl, HDL < 50 mg/dl in women and 40 mg/dl in men, triglycerides >150 mg/dl;
- family history: cardiovascular symptoms in the first-degree relatives (myocardial infarction, sudden death, or coronary disease) in women less than 65 years and men less than 55 years of age.

2.1 COPD and Ischemic Heart Disease (IHD) Classifications

The severity of COPD was assessed according to the Global Initiative for Chronic Obstructive Lung Disease criteria (GOLD 2016). The five-point modified Medical Research Council (mMRC) scale of dyspnea was used to assess the ability of patients to perform daily activities (Perez et al. 2015).

The evaluation of IHD stability was performed using the classification of Canadian Cardiovascular Society (Campeau 2002; Dagenais et al. 2002; Campeau 1976). Anginal pain at rest may be present in each class of patients, resulting from the coexisting and overlapping contraction of coronary vessels. The classes outlined below concern the maximum limitation, and patients' function may be better and variable from day-to-day.

- Class I: average daily activities cause no anginal pain. Pain occurs on intense fast or longer-lasting exertion;
- Class II: slight limitation of average activities. Anginal pain occurs during fast walking, climbing stairs, activity performed after a meal, in low surrounding temperature, during emotional stress, within the first few hours after waking up. Patients can walk distances on a flat area, climb more than one flight of stairs;
- Class III: Patients experience some limitation to average daily activities. Anginal pain occurs on walking distances of several kilometres on flat area or climbing one flight of stairs;
- Class IV: Patients experience limitations to performing any physical activity. Angina symptoms may develop at rest.

2.2 Medication Expenditures

The estimate of medication expenditure for IHD treatment was performed as based on medical records of patients. Retail cost of medications was calculated based on the list published in 2016 by the Minister of Health in Poland. Unit costs were taken from the appraisement of the Polish National Health Fund. A 30-day time scale was assumed in the evaluation. Since we specifically focused on the estimation of costs for heart disease treatment, the costs associated with COPD treatment were omitted.

This study had a nonexperimental design where we examined and compared by gender medication expenditures against the benefits gained during treatment of IHD in the course of COPD, in a natural setting without any manipulation or control. We identified a group of patients with IHD during the course of COPD, taking a particular IHD-specific drug treatment, followed them over time, and then compared the medication spending against health outcomes, expressed as the number and percentage of patients displaying reductions in IHD risk factors. This collection of reports on the drug treatment of individual patients, multiplicity of drugs assessed, widely different drug retail costs, different levels of rebate by the National Health Fund, and a relatively small number of patients in each severity class of IHD made specific statistical data elaboration unreasonable. Changes in the number of patients still displaying specific risk factors for IHD after 2-month-long treatment *versus* before treatment were compared with a Chi-square 2 × 2 contingency tables for categorical data. A p-value less than 0.05 defined statistically significant differences.

3 Results and Discussion

All patients studied were allocated to COPD category C, i.e., a higher risk category due to FEV1 lower than 50% predicted or higher frequency of exacerbations, based on the GOLD (2016) criteria; having the mean FEV1 of 0.90 ± 0.18 L/s and mMRC dyspnea score of 2.7 ± 0.5. They were evaluated twice: on the day of first examination before treatment and after 2 months of treatment (Table 1). Among the risk factors for IHD, cigarette smoking, history of hyperlipidemia, and lack of physical activity were most prevalent in both men and women

Table 1 Changes in the number (%) of patients, before *versus* after 2 months of treatment, displaying demographic and clinical features of ischemic heart disease (IHD) accompanying chronic obstructive pulmonary disease (COPD)

Features	Before treatment		After treatment	
	Women (*n* = 27)	Men (*n* = 30)	Women (*n* = 27)	Men (*n* = 30)
Age; years (range)	46 (41–49)	47 (43–47)	46 (41–49)	47 (43–47)
BMI; kg/m^2 (range)	30 (27–33)	28 (24–30)	26 (24–30)	27 (24–30)
Physical activity; n (%)	19 (70)	14 (47)	24 (89)	16 (53)*
Alcohol use; n (%)	9 (33)	17 (57)	3 (11)**	15 (50)*
Cigarette smoking; n (%)	26 (96)	27 (90)	19 (70)**	24 (80)
Diabetes; n (%)	2 (7)	4 (13)	2 (7)	4 (13)
Arterial hypertension; n (%)	10 (37)	12 (30)	8 (30)	11 (37)
Hyperlipidemia; n (%)	20 (74)	27 (90)	15 (55)	25 (83)*
Relevant family history; n (%)	12 (44)	15 (50)	12 (44)	15 (50)

*$p \leq 0.02$ for inter-gender differences in a given treatment group; **$p \leq 0.05$ for the corresponding feature in the same gender after treatment

before treatment, with a greater number of alcohol and smoking users among men than women. In contradistinction, women were more physically active than men. After 2 month-long IHD treatment, improper physical activity remained an issue in both gender groups, with women being less physically active both before and after treatment, while the number of patients exhibiting the other risk factors decreased. There was a significant decrease in the number of women using alcohol and cigarettes after treatment *versus* before treatment. Despite the decreases, alcohol use and hyperlipidemia remained significantly more prevalent in men than women after the treatment. Comparison of the number and percentage of patients having serum biochemical indices confirmed the prevalence of lipid and glucose dysmetabolism in IHD – COPD patients. The prevalence of dysmetabolic patients ranged between 70 and 90% in both gender groups before treatment and it decreased to about 50–80% after treatment. Hyperlipidemia remained prevalent in men after treatment, as its frequency significantly decreased in women (Table 2). Nonetheless, improvements speak for a better control of lipid and glucose metabolism due to the pharmacotherapy employed.

Pharmacotherapy costs were evaluated on the basis of medical records of all 57 patients, gender combined, treated from June to December in 2016, with the medication prices being then in effect. Medication-specific and total treatment costs stratified by IHD class one-year-long treatment are displayed in Table 3. The table shows the retail price of drugs and compares that against various drug-specific cost-refunds routinely applied by the Polish National Health Fund, being in force during the study time. There is an apparent upward leap in total medication spending between Class I-II and Class III-IV of IHD. *Per capita per year* retail spending increased with IHD severity, being outstandingly greater in the most severe class compared to the mildest IHD class; 188.80 EUR and 29.71 EUR, respectively, which was almost a 6.4-fold increase.

Refunds applied by the Polish National Health Fund slashed the treatment costs by about 47% in Class I, 44% in Class II, 26% in Class III, and 52% in Class IV of IHD (Table 3, Fig. 1).

Cardiovascular diseases and COPD are global epidemics incurring significant morbidity and mortality. According to WHO (2011), heart diseases mortality rate will soon reach 20 million *per* year. IHD most commonly affects men in productive age, which has a serious economic bearing (Czech et al. 2013). IHD also is a frequently accompaniment of COPD. The prognosis for COPD combined with heart disease is poorer than that for either disease alone. In this study, several typical risk factors for IHD were observed in the course of COPD, i.e., hyperlipidemia, arterial hypertension, obesity, diabetes, low physical activity, tobacco smoking, alcohol abuse, and family history (European Association for

Table 2 Changes in the number (%) of patients with biochemical signs of lipid dysmetabolism, before *versus* after 2 months of treatment, as risk factors for ischemic heart disease (IHD) accompanying chronic obstructive pulmonary disease (COPD)

Biochemical indices	Before treatment n (%)		After treatment n (%)	
	Women (*n* = *27*)	Men (*n* = *30*)	Women (*n* = *27*)	Men (*n* = *30*)
Total cholesterol >200 mg/dl	20 (74)	27 (90)	15 (55)	25 (83)*
LDL >130 mg/dl	24 (89)	28 (93)	14 (52)**	25 (83)*
HDL <40 mg/dl	23 (85)	26 (87)	13 (48)**	24 (80)*
Triglycerides >180 mg/dl	25 (93)	28 (93)	14 (52)**	24 (80)*
Glucose >110 mg/dl	19 (70)	25 (83)	13 (48)	22 (73)

*$p \leq 0.02$ for inter-gender differences in a given treatment group; **$p \leq 0.004$ for the corresponding feature in the same gender after treatment

Table 3 Summary *per capita per year* costs of medications (EUR) for treatment of ischemic heart disease (IHD, stratified into four classes of severity, accompanying chronic obstructive pulmonary disease (COPD)

Generic medication name	Payment	Patients taking a drug (n)	One-year treatment	
			Retail price	Discounted price[a]
Class I – 14 patients				
Acetylsalicylic acid; 75 mg/day	100%	14	9.57	9.57
Nitroglycerin; 1 dose	Lump-sum	6	3.09	0.60
Bisoprolol; 10 mg/day	Lump-sum	8	17.05	5.61
	Total/patient (EUR)		29.71	15.78
Class II – 13 patients				
Acetylsalicylic acid; 75 mg/day	100%	13	8.89	8.89
Nitroglycerin; 1 dose	Lump-sum	7	3.61	0.70
Bisoprolol; 10 mg/day	Lump-sum	5	10.66	3.51
	Total/patient (EUR)		23.16	13.10
Class III – 12 patients				
Acetylsalicylic acid; 150 mg/day	100%	12	17.08	17.08
Nitroglycerin; 2 doses	Lump-sum	5	6.02	1.20
Perindopril; 5 mg/day	Lump-sum	1	5.20	4.51
Metoprolol; 100 mg/day	Lump-sum	7	13.87	7.78
Simvastatin; 40 mg/day	30%	10	69.13	51.74
	Total/patient (EUR)		111.30	82.31
Class IV – 8 patients				
Acetylsalicylic acid; 150 mg/day	100%	8	17.08	17.08
Nitroglycerin; 2 doses	Lump-sum	8	14.46	2.88
Ramipril; 10 mg/day	Lump-sum	8	58.66	34.98
Atrovasterol; 80 mg/day	30%	8	98.60	35.11
	Total/patient (EUR)		188.80	90.05

[a]After a routine rebate applied by the Polish National Health Fund, being in force during the study time

Cardiovascular Prevention and Rehabilitation et al. 2011; Collet et al. 2007; Leitzmann et al. 2007; Warburton et al. 2006). Patients surveyed in this study used antiplatelet medications, β-adrenolytics, renin-angiotensin-aldosterone inhibitors, and statins, all of which are essential in the prevention of death risk or acute cardiovascular incidents. Davies et al. (2010) have shown a decrease in death risk by 20% and in myocardial infarction by 41% in IHD patients on antiplatelet acetylsalicylic acid therapy. In the same study, β-adrenolytics reduce the risk of developing acute cardiovascular incidents by 29%, statins reduce the risk of death by 18%

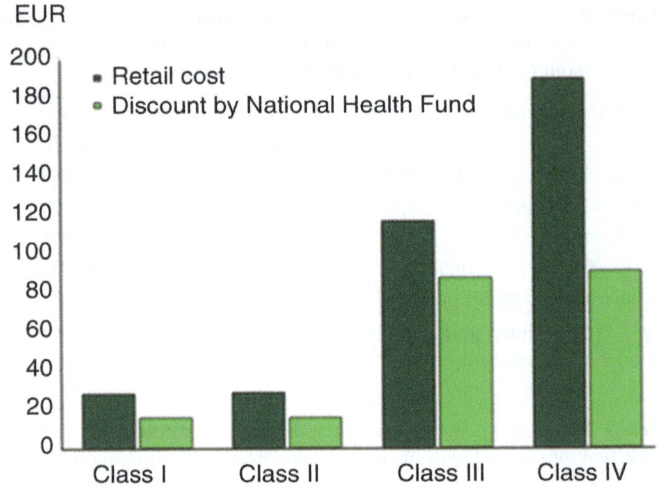

Fig. 1 Spending for medications (EUR) *per capita per year* for treatment of ischemic heart disease (IHD), stratified into four classes of severity, accompanying chronic obstructive pulmonary disease (COPD)

and coronary incidents by 31%, and angiotensin converting enzyme inhibitors considerably improve the prognosis in heart failure.

In the present study, we found a considerable fivefold increase in *per capita* spending in advanced IHD accompanying COPD. This spending does not include the treatment for the basic disease, i.e., COPD. The seriousness of the economic barrier for treatment has been recognized by the National Health Fund that offers almost a 50% cost refund in IHD-COPD pathology. We also considered a need to control for non-pharmacological factors, such as pro-health education and avoidance of alcohol or tobacco addiction, which play a role in reducing the risk of developing cardiovascular incidents, and thus could decrease the spending linked to pharmacotherapy (Thirapatarapong et al. 2014). It seems, however, that these sociobehavioral factors were of no avail in closing the spending gap for medications between the mild and the severe IHD in COPD. Taken together these findings underscore the role of an early diagnosis of heart-related comorbidities in the course of COPD for the appropriate treatment cost allocations. This study also reaffirms that heart disease accompanying COPD is a frequent occurrence, generating high costs of treatment, which relates to the severity of this comorbidity.

Conflicts of Interest The authors declare no conflicts of interest in relation to this article.

References

Brekke PH, Omland T, Smith P, Søyseth V (2008) Underdiagnosis of myocardial infarction in COPD – Cardiac Infarction Injury Score (CIIS) in patients hospitalised for COPD exacerbation. Respir Med 102 (9):1243–1247

Buist AS, MA MB, Vollmer WM, Gillespie S, Burney P, Mannino DM, Menezes AM, Sullivan SD, Lee TA, Weiss KB, Jensen RL, Marks GB, Gulsvik A, Nizankowska-Mogilnicka E, BOLD Collaborative Research Group (2007) International variation in the prevalence of COPD (the BOLD study): a population-based prevalence study. Lancet 370(9589):741–750

Bursi F, Vassallo R, Weston SA, Killian JM, Roger VL (2010) Chronic obstructive pulmonary disease after myocardial infarction in the community. Am Heart J 160(1):95–101

Campeau L (1976) Letter: grading of angina pectoris. Circulation 54(3):522–523

Campeau L (2002) The Canadian cardiovascular society grading of angina pectoris revisited 30 years later. Can J Cardiol 18(4):371–379

Campo G, Guastaroba P, Marzocchi A, Santarelli A, Varani E, Vignali L, Sangiorgio P, Tondi S, Serenelli C, De Palma R, Saia F (2013) Impact of COPD on long-term outcome after ST-segment elevation myocardial infarction receiving primary percutaneous coronary intervention. Chest 144(3):750–757

Chatila WM, Thomashow BM, Minai OA, Criner GJ, Make BJ (2008) Comorbidities in chronic obstructive pulmonary disease. Proc Am Thorac Soc 5(4):549–555

Collet F, Mallart A, Bervar JF, Bautin N, Matran R, Pattou F, Romon M, Perez T (2007) Physiologic correlates of dyspnea in patients with morbid obesity. Int J Obes 31(4):700–706

Curkendall SM, DeLuise C, Jones JK, Lanes S, Stang MR, Goehring E Jr, She D (2006) Cardiovascular disease in patients with chronic obstructive pulmonary disease, Saskatchewan Canada cardiovascular disease in COPD patients. Ann Epidemiol 16(1):63–70

Czech M, Opolski G, Zdrojewski T (2013) The costs of heart failure in Poland from the public payer's perspective. Polish programme assessing diagnostic procedures, treatment and costs in patients with heart failure in randomly selected outpatient clinics and hospitals at different levels of care: POLKARD. Kardiol Pol 71:224–232

Dagenais GR, Armstrong PW, Théroux P, Naylor CD, CCS Ad Hoc Committee for Revising the CCS Grading of Stable Angina (2002) Revisiting the Canadian Cardiovascular Society grading of stable angina pectoris after a quarter of a century of use. Can J Cardiol 18(9):941–944

Davies EJ, Moxham T, Rees K, Singh S, Coats AJ, Ebrahim S, Lough F, Taylor RS (2010) Exercise training for systolic heart failure: Cochrane systematic review and meta-analysis. Eur J Heart Fail 12 (7):706–715

Enriquez JR, Parikh SV, Selzer F, Jacobs AK, Marroquin O, Mulukutla S, Srinivas V, Holper EM (2011) Increased adverse events after percutaneous coronary intervention in patients with COPD: insights from the National Heart, Lung, and Blood Institute dynamic registry. Chest 140:604–610

European Association for Cardiovascular Prevention & Rehabilitation, Reiner Z, Catapano AL, De Backer G, Graham I, Taskinen MR, Wiklund O, Agewall S, Alegria E, Chapman MJ, Durrington P, Erdine S, Halcox J, Hobbs R, Kjekshus J, Filardi PP, Riccardi G, Storey RF, Wood D, ESC Committee for Practice Guidelines (CPG) 2008–2010 and 2010–2012 Committees (2011) ESC/EAS guidelines for the management of dyslipidaemias: the task force for the management of dyslipidaemias of the European Society of Cardiology (ESC) and the European Atherosclerosis Society (EAS). Eur Heart J 32(14):1769–1818

Fabbri LM, Luppi F, Beghe B, Rabe KF (2008) Complex chronic comorbidities of COPD. Eur Respir J 31 (1):204–212

Fear JR, Rodrigues LC, Smith CJ, Hubbard RB, Gibson JE (2010) Prevalence of major comorbidities in subjects with COPD and incidence of myocardial infarction and stroke: a comprehensive analysis using data from primary care. Thorax 65(11):956–962

GOLD (2016) From the global strategy for the diagnosis, management and prevention of COPD, Global initiative for chronic obstructive lung disease. http://goldcopd.org/global-strategy-diagnosis-management-prevention-copd-2016/. Accessed on 30 Jan 2018

Gudmundsson G, Gislason T, Lindberg E, Hallin R, Ulrik CS, Brøndum E, Nieminen MM, Aine T, Bakke P, Janson C (2006) Mortality in COPD patients discharged from hospital: the role of treatment and co-morbidity. Respir Res 7:109

Leitzmann MF, Park Y, Blair A, Ballard-Barbash R, Mouw T, Hollenbeck AR, Schatzkin A (2007) Physical activity recommendations and decreased risk of mortality. Arch Intern Med 167(22):2453–2460

Mannino DM, Thorn D, Swensen A, Holguin F (2008) Prevalence and outcomes of diabetes, hypertension and cardiovascular disease in COPD. Eur Respir J 32 (4):962–969

Perez T, Burgel P, Paillasseur J, Caillaud D, Deslée G, Chanez P, Roche N, Scientific Committee INITIATIVESBPCO (2015) Modified Medical Research Council scale vs baseline dyspnea index to evaluate dyspnea in chronic obstructive pulmonary disease. Int J Chron Obstruct Pulmon Dis 10:1663–1672

Sin DD, Man SF (2005) Chronic obstructive pulmonary disease as a risk factor for cardiovascular morbidity and mortality. Proc Am Thorac Soc 2(1):8–11

Soriano JB, Visick GT, Muellerova H, Payvandi N, Hansell AL (2005) Patterns of comorbidities in newly diagnosed COPD and asthma in primary care. Chest 128(4):2099–2107

Soriano JB, Rigo F, Guerrero D, Yañez A, Forteza JF, Frontera G, Togores B, Agustí A (2010) High prevalence of undiagnosed airflow limitation in patients with cardiovascular disease. Chest 137(2):333–340

Thirapatarapong W, Armstrong HF, Bartels MN (2014) Comparison of cardiopulmonary exercise testing variables in COPD patients with and without coronary artery disease. Heart Lung 43(2):146–151

van Manen JG, Bindels PJ, Ijzermans CJ, van der Zee JS, Bottema BJAM, Schadé E (2001) Prevalence of comorbidity in patients with a chronic airway obstruction and controls over the age of 40. J Clin Epidemiol 54(3):287–293

Warburton DE, Nicol CW, Bredin SS (2006) Health benefits of physical activity: the evidence. CMAJ 174 (6):801–809

WHO (2011) World Health Organization, Geneva. Available from http://www.who.int/gho/publications/world_health_statistics/EN_WHS2011_Full.pdf. Accessed on 20 Dec 2017

Yusuf S, Hawken S, Ounpun S, Dans T, Avezum A, Lanas F, McQueen M, Budaj A, Pais P, Varigos J, Lisheng L, Study Investigators INTERHEART (2004) Effect of potentially modifiable risk factors associated with myocardial infarction in 52 countries (the INTERHEART study): case-control study. Lancet 364(9438):937–952

Zdrojewski T, Rutkowski M, Bandosz P, Gaciong Z, Jędrzejczyk T, Solnica B, Pencina M, Drygas W, Wojtyniak B, Grodzicki T, Piwoński J, Wyrzykowski B (2013) Prevalence and control of cardiovascular risk factors in Poland. Assumptions and objectives of the NATPOL 2011 survey. Kardiol Pol 71(4):381–392

Advs Exp. Medicine, Biology - Neuroscience and Respiration (2018) 41: 57–66
https://doi.org/10.1007/5584_2018_199

Published online: 21 April 2018

Proangiogenic and Profibrotic Markers in Pulmonary Sarcoidosis

I. Tuleta, L. Biener, C. Pizarro, G. Nickenig, and D. Skowasch

Abstract

The aim of our study was to determine the blood levels of vascular endothelial growth factor (VEGF), transforming growth factor (TGF)-β1, fibroblast growth factor (FGF)-2, and platelet-derived growth factor (PDGF)-AB in different stages of pulmonary sarcoidosis. There were 92 patients in sarcoidosis stages I + II, III, and IV enrolled into the study. All the patients underwent lung diffusing capacity and blood sampling. We found that VEGF levels differed significantly between the stage groups with the peak VEGF concentrations in stage III. TGF-β1 levels were similar in stages I + II and III, and tended to be lower in stage IV. The analysis of the subgroups showed increased VEGF and FGF-2, and reduced TGF-β1 concentration in stages I + II patients with relevantly reduced lung diffusing capacity or increased sarcoidosis activity compared to patients with normal lung diffusing capacity or inactive sarcoidosis. A tendency towards increased VEGF, PDGF-AB and TGF-β1 levels was observed in the analogical subgroup analysis within the stage III. We conclude that proangiogenic VEGF, and profibrotic FGF-2 and PDGF-AB may contribute to the progression of sarcoidosis, whereas TGF-β1, with its dual anti-inflammatory and profibrotic actions, may play a dichotomous protective or deleterious role. Reduced diffusing capacity and active sarcoidosis are associated with an unfavorable constellation of the markers studied, which predicts a progressive disease course.

Keywords

Angiogenesis · Fibrosis · Granulomatous disorder · Inflammation · Markers · Matrix accumulation · Parenchymal disease · Pulmonary sarcoidosis · Vascular permeability

I. Tuleta (✉), L. Biener, C. Pizarro, G. Nickenig, and D. Skowasch
Department of Internal Medicine II – Cardiology, Pulmonology and Angiology, University of Bonn, Bonn, Germany
e-mail: izat@gmx.de

1 Introduction

Sarcoidosis is an inflammatory granulomatous disease which may potentially affect each organ with a predilection for pulmonary tissue (Newman et al. 1997). The course of sarcoidosis is highly variable: most of the patients undergo remission, the others develop chronic sarcoidosis. Pulmonary fibrosis in frame of sarcoidosis progression represents the end stage organ damage (Iannuzzi and Fontana 2011). Pulmonary fibrosis affects approximately 20% of patients and is associated with increased morbidity and mortality (Nardi et al. 2011; Swigris et al. 2011; Baughman et al. 1997). The pathomechanisms of sarcoidosis progression to the irreversible stage of pulmonary fibrosis are not clear. There are no data confirming that the treatment of sarcoidosis at its very early stage or application of immunosuppressive drugs may prevent or change disease progression (King et al. 2001; Judson 1999). Different markers and cytokines such as Th2 cytokines (Hauber et al. 2003), interleukin (IL)-7 (Patterson et al. 2013), IL-4, IL-9, IL-10, IL-13, and transforming growth factor (TGF)-β (Noor and Knox 2007) have been demonstrated to be increased in sarcoidosis-related pulmonary fibrosis. However, it is little known about the predictive factors for sarcoidosis progression.

Therefore, the aim of this study was to determine the serum levels of proangiogenic and profibrotic markers, such as: vascular endothelial growth factor (VEGF), TGF-β1, fibroblast growth factor (FGF)-2 and platelet-derived growth factor (PDGF)-AB in different stages of pulmonary sarcoidosis and to correlate them with extrapulmonary manifestation, cortisone therapy, lung diffusing capacity, and sarcoidosis activity.

2 Methods

The research was conducted according to the principles of the Declaration of Helsinki and was approved by the local Ethics Committees. Written informed consent was obtained from all the patients.

2.1 Patients

Ninety two outpatients with pulmonary sarcoidosis in stages I + II, III, and IV, with or without extrapulmonary involvement, visiting our department of pneumology at the University of Bonn between May and December 2015 were enrolled in the study. The diagnosis of sarcoidosis was biopsy-proven by means of bronchoscopy. Exclusion criteria comprised other relevant pulmonary diseases, such as chronic obstructive pulmonary disease (COPD), asthma, lung tumor, or pulmonary fibrosis not in frame of sarcoidosis, as well as disorders involving chronic inflammation or fibrotic processes. All patients underwent the measurements of diffusing capacity of the lung for carbon monoxide (DLCO) and blood sampling.

2.2 Measurements

Lung Diffusing Capacity DLCO in a single breath method was assessed by body plethysmography (Jaeger©, Alveo-Diffusionstest Jaeger©, Wuppertal, Germany). Predicted values were calculated automatically based on the age, sex, height, and race by the software of the pulmonary function test. Percentages of the predicted values for DLCO were used for the subsequent statistical analysis.

Blood Tests Sarcoidosis activity markers, such as: angiotensin converting enzyme (ACE) and soluble interleukin-2 receptor (sIL-2R) were determined in the serum. Beyond that, VEGF, TGF-β1, FGF-2 and PDGF-AB serum levels were measured by an enzyme-linked immunosorbent assay (ELISA).

2.3 Statistical Analysis

Parametric values were expressed as means ±SD. Differences between the means of three groups were calculated by ANOVA and between two groups by Student's *t*-test. Comparisons between categorical variables were performed by Chi-square test. A p-value <0.05 defined statistically significant differences. The analysis was conducted using a commercial statistical package of SPSS v23 (IBM Corp., Armonk, NY).

3 Results

Sarcoidosis patients were divided into three groups according to the radiological stage of pulmonary sarcoidosis: group 1 in stages I + II, group 2 in stage III, group 3 in stage IV, as presented in the Table 1. There were no relevant differences in demographics or the frequency of

Table 1 Baseline characteristics of sarcoidosis patients in different stages of pulmonary disease

Sarcoidosis	Sex F/M	Age (year)	BMI (kg/m^2)	Cortisone therapy n (%)	Additional extra-pulmonary involvement n (%)	Increased ACE and /or sIL-2R# n (%)	Reduction in DLCO n (%)			
							No	Low	Moderate	Severe
Group 1; stage I + II	37/30	51.9 ± 11.9	27.9 ± 5.7	28 (41.8)	23 (34.3)	29 (43.3)	31 (46.3)	20 (29.9)	14 (20.9)	2 (3.0)
Group 2; stage III	12/8	58.5 ± 12.5	26.3 ± 5.5	16 (80.0)	9 (45.0)	8 (40.0)	5 (25)	8 (40)	4 (20)	3 (15)
Group 3; stage IV	3/2	53.2 ± 12.3	24.2 ± 3.0	5 (100)*	2 (40)	3 (60)	0 (0)	1 (20)	2 (40)	2 (40)*

There were no differences in sex, age, BMI, or the percentage of patients with increased ACE and/or sIL-2R between the groups. Along with disease progression significantly more patients were under cortisone therapy and presented reduced DLCO

F/M female/male, *BMI* body mass index, *ACE* angiotensin converting enzyme, *sIL-2R* soluble interleukin-2 receptor, *DLCO* lung diffusion capacity for carbon monoxide. Results are means ±SE or n (% of study patients); *$p < 0.05$; #increased ACE >85 U/L, increased sIL-2R >730 U/mL. Reduction in lung diffusing capacity: no ≥80% of predicted, low ≥65 < 80% of predicted, moderate ≥50 < 65% of predicted, severe <50% of predicted additional extrapulmonary disease manifestation between the groups. As the disease progressed, DLCO significantly decreased and more patients were on cortisone therapy. The activity of sarcoidosis defined as increased blood levels of ACE of ≥85 U/L and/or sIL-2R ≥730 U/mL did not differ across the groups. There were significant differences in VEGF concentration in the blood across the patient groups studied ($p = 0.027$) (Fig. 1a). Specifically, VEGF level increased along with the worsening stage of the disease until pulmonary stage III ($p = 0.009$ *vs.* stages I + II). Pulmonary sarcoidosis in stage IV was characterized by a decreased VEGF blood concentration compared to that in stage III; although the difference failed to reach significance ($p = 0.388$). In contradistinction, TGF-β1 levels were similar for the pulmonary stages I + II and III (Fig. 1b). Analogous to VEGF, TGF-ß1 tended to be lower in stage IV. No differences in the blood levels of FGF-2 or PDGF-AB were detected between the groups (Fig. 1c and d).

To assess which factors or clinical conditions are associated with different serum levels of the four markers studied, subgroups of patients within each of three groups were created depending on the presence of concomitant extrapulmonary manifestations, cortisone therapy, reduction in DLCO, and sarcoidosis activity. In group 1, i.e., disease stage I + II, there was no

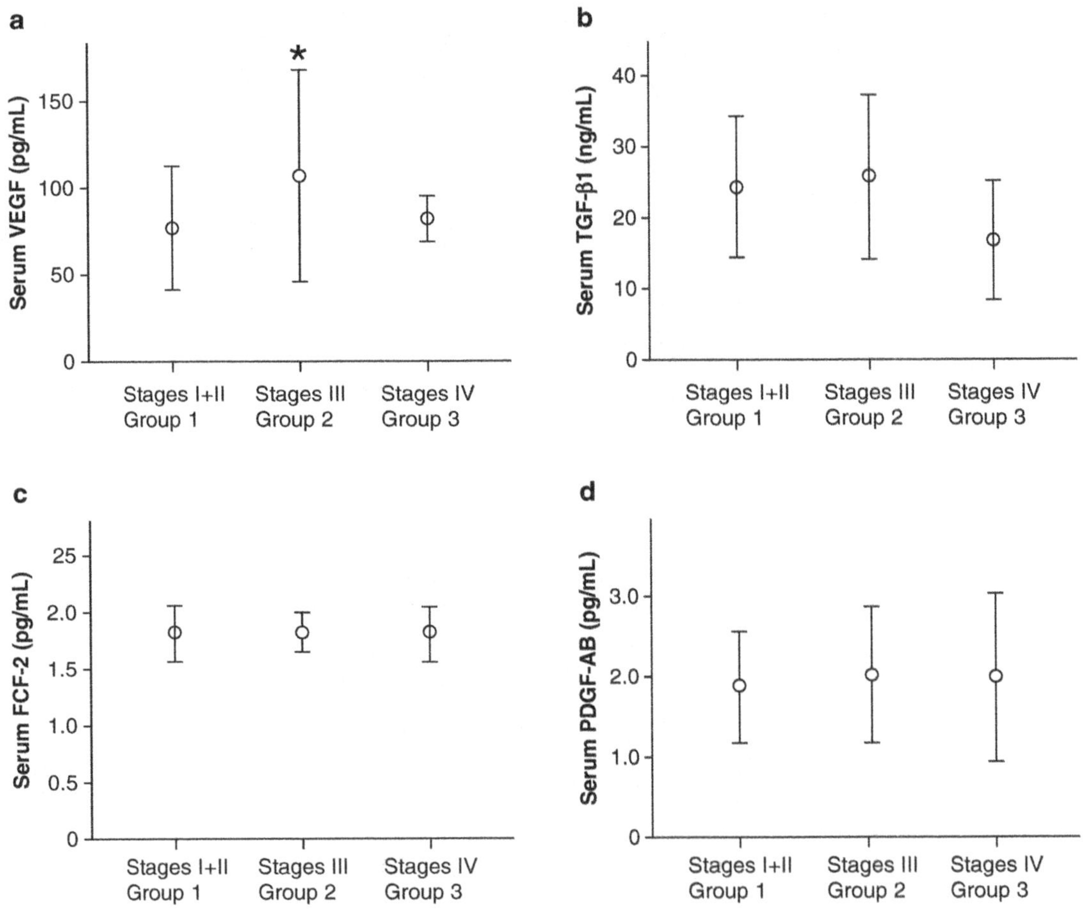

Fig. 1 Serum content of vascular endothelial growth factor (VEGF), transforming growth factor (TGF)-β1, fibroblast growth factor (FGF)-2, and platelet-derived growth factor (PDGF)-AB in different sarcoidosis stages. Data are means ±SD; *$p < 0.05$

correlation between additional extrapulmonary involvement or cortisone therapy and the levels of the markers examined. Moderate-to-severe reduction in DLCO (n = 16) was related to a significant increase in VEGF (94.0 ±49.7 pg/mL) and FGF-2 (1.9 ±0.2 pg/mL), and a tendency toward lower TGF-β1 levels (21.9 ±8.3 ng/mL) compared to the patients with no or only mild reduction in DLCO (n = 51) (72.8 ±27.7 pg/mL, p = 0.036; 1.8 ±0.3 pg/mL, p = 0.043; 25.2 ± 10.3 ng/mL, p = 0.246, respectively for VEGF, FGF-2, and TGF-β1) (Fig. 2a). No differences were observed in PDGF-AB levels in reduced *vs.* non-reduced DLCO subgroups (1.9 ±0.7 pg/mL *vs.* 1.9 ±0.7 pg/mL, respectively). In patients with active pulmonary sarcoidosis, no relevant alternations in VEGF, TGF-β1, FGF-2, and PDGF-AB serum levels were detected compared to those in individuals with inactive sarcoidosis. However, as we set the cut-off values lower for ACE (52 U/L) and higher for sIL-2R (900 U/mL) according to other studies (Rothe et al. 2010), significant decreases in TGF-β1 were detected in both patient collectives when either ACE or sIL-2R were elevated. TGF-β1 amounted to 20.92 ±9.01 ng/mL (Fig. 2b) in case of elevated ACE and to 19.34 ±9.29 ng/mL (Fig. 2c) in case of elevated sIL-2R, compared to the 26.36 ±9.97 ng/mL and 26.04 ±9.69 ng/mL values without the ACE/sIL-2R elevations in blood, respectively.

In group 2, i.e., disease stage III, there were only trends toward higher serum levels of VEGF (144.5 ±83.2 pg/mL) and PDGF-AB (2.5 ±1.2 pg/mL) in patients with a relevant reduction in DLCO *vs.* those with no or only mild DLCO reduction (93.8 ±42.6 pg/mL and 1.8 ±0.6 pg/mL, respectively) (Fig. 3a). There were no relevant differences in the levels of the markers studied depending on the ACE blood concentration (Fig. 3b). Patients in this group, with sIL-2R values ≥900 U/mL, presented a strong trend toward higher VEGF (149.8 ±89.4 pg/mL) (Fig. 3c) and TGF-β1 (34.0 ±7.8 ng/mL) (Fig. 3c) in the blood compared to those with sIL-2R <900 U/mL (92.7 ± 42.9 pg/mL, p = 0.066 and 22.7 ±11.3 ng/mL, p = 0.059, respectively).

In group 3, i.e., disease stage IV, no evaluation could be conducted due to a small number of patients.

4 Discussion

This study examined serum markers involved in inflammation and fibrosis in different stages of sarcoidosis. The findings demonstrate that serum VEGF content increased in stage III of sarcoidosis compared to that present in the early disease stages. However, in the most advanced stage IV, VEGF fell back to the level of the early stages. Sekiya et al. (2003) have reported a higher serum VEGF level in patients with severe sarcoidosis who received corticosteroid treatment. Fireman et al. (2009) have reported a lower VEGF level in induced sputum, collected from sarcoidosis patients being in stages III and IV than that in the stages I and II. Other studies failed to report any relevant differences in the blood levels of this marker in sarcoidosis, depending on the extent of parenchymal involvement, although these observations are hampered by a non-enrollment of stage IV patients in those studies (Piotrowski et al. 2015; Ziora et al. 2015). VEGF is a pleiotropic cytokine which plays an important role in the angiogenesis, vascular permeability, and inflammation (Tolnay et al. 1997). It may be produced by monocytes in granulomas in sarcoidosis (Yamashita et al. 2013). Granulomas predominate in the lung tissue of sarcoidosis patients in stage III, which may account for the augmented VEGF concentration in stage III found in the present study. Likewise, reduction in VEGF in stage IV demonstrated in our work could be explicable by a gradual disappearance of granulomas and their replacement by fibrotic tissue.

In contradistinction to VEGF changes, there were no significant differences between serum TGF-β1 levels in stages I + II, and III of sarcoidosis. However, there was a tendency toward lower TGF-β1 content in stage IV compared to other stages. These results are in line with those of Piotrowski et al. (2015) who failed to notice any differences in TGF-β1 between stages I-III.

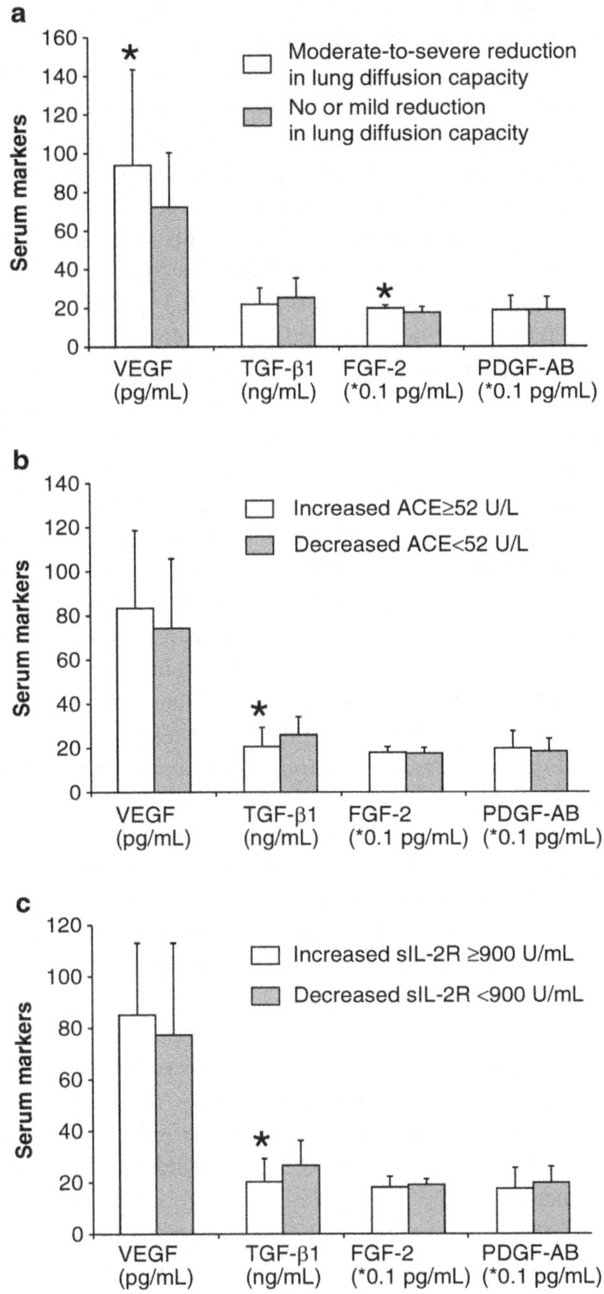

Fig. 2 Different serum content of vascular endothelial growth factor (VEGF), transforming growth factor (TGF)-β1, fibroblast growth factor (FGF)-2, and platelet-derived growth factor (PDGF)-AB according to the lung diffusing capacity (**a**), angiotensin converting enzyme (ACE) (**b**), and soluble interleukin-2 receptor (sIL-2R) (**c**) in the patient serum in stage I + II of sarcoidosis. Moderate-to-severe reduction in lung diffusing capacity = <65% of the predicted value, $*p < 0.05$

TGF-β may modulate inflammatory processes and has a profibrotic action (Marie et al. 2006). Among three TGF-β isoforms, TGF-β1 is notably involved in the fibrotic process in sarcoidosis (Willis et al. 2005). TGF-β has been detected in granulomas from patients with skin manifestation of sarcoidosis (Marshall et al. 1996) and in granulomatous lesions in pulmonary sarcoidosis (Limper et al. 1994). Akin to VEGF, reduction in granuloma size in favor of matrix accumulation, resulting in scar formation, could partly explain a lower TGF-β1 concentration in the

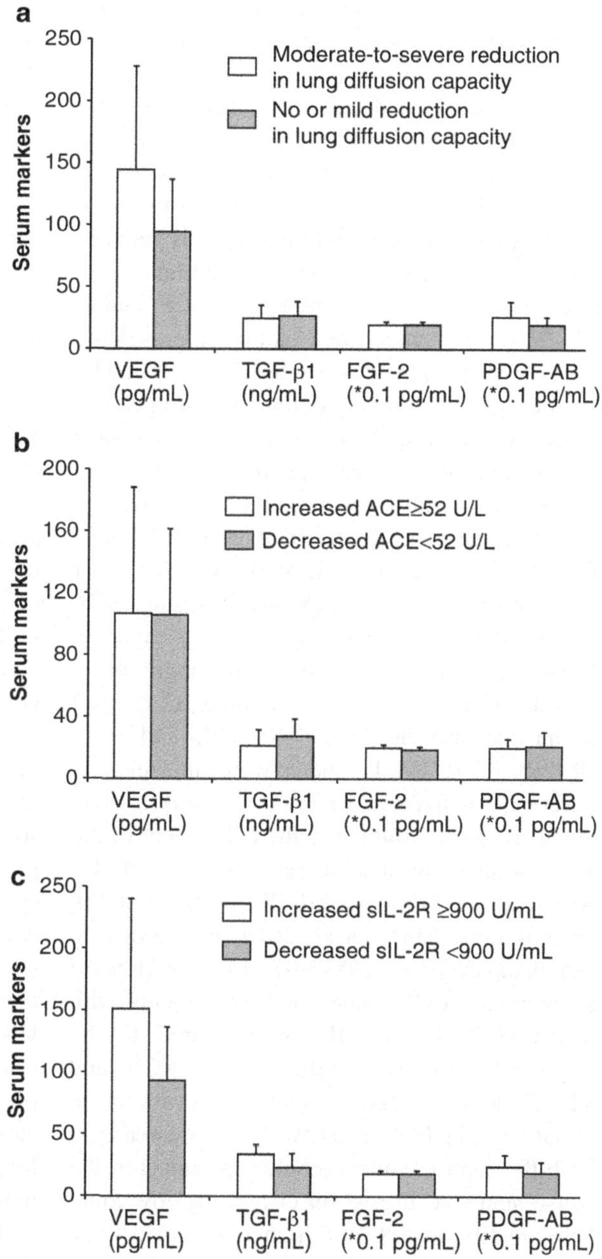

Fig. 3 Different serum content of vascular endothelial growth factor (VEGF), transforming growth factor (TGF)-β1, fibroblast growth factor (FGF)-2, and platelet-derived growth factor (PDGF)-AB according to the lung diffusing capacity (**a**), angiotensin converting enzyme (ACE) (**b**), and soluble interleukin-2 receptor (sIL-2R) (**c**) in the patient serum in stage III of sarcoidosis. Moderate-to-severe reduction in lung diffusing capacity = <65% of the predicted value

irreversible stage IV of sarcoidosis we observed in this study. Interestingly, higher TGF-β levels were associated with spontaneous regression of sarcoidosis (Zissel et al. 1996). Further, Hauber et al. (2003) have noticed that IL-13, which stimulates the release of TGF-β, was increased in patients without parenchymal disease, again however patients in stage IV were not included in that study. Additionally, polymorphic alleles of TGF-β1, associated with low levels of this protein production, were linked to more severe disease presentation (Jonth et al. 2007).

Prognosis of sarcoidosis depends mainly on the disease progression which, in turn, is determined by the interplay between various factors. The examined markers related to the stages of sarcoidosis represent only a snapshot in the course of the disease. It seems much more important to define which patients are prone to develop a progressive form of sarcoidosis. Therefore, the above mentioned markers were compared between patients with a relevantly reduced lung diffusing capacity or pronounced inflammation, expressed as an increase in ACE and/or sIL-2R serum content, and patients representing the same stage of the disease, but characterized by no or only mild reduction in lung diffusing capacity or lower ACE and sIL-2R content. The results demonstrate that patients with sarcoidosis in stages I + II, with more impaired lung diffusing capacity, had significantly increased VEGF and FGF-2 levels and they tended to have lower TGF-β1 levels compared to the same stage subgroup but with a better lung diffusing capacity. When comparing the stage I + II subgroups in regard to the activity of sarcoidosis, defined as an increase in ACE ≥85 U/L and/or in sIL-2R ≥730 U/mL, there were no relevant differences in the level of any of the four markers tested. However, following the adjustment of the upper limit of normal range to the 52 U/L for ACE and 900 U/mL for sIL-2R, as described in other reports (Rothe et al. 2010), we noticed a significant decrease in TGF-β1 and a trend toward augmented VEGF values in both patients with increased ACE and sIL-2R compared to the respective subgroups with normal ACE and sIL-2R levels. Taken together, elevations in VEGF and in FGF-2, as well as a decrease in TGF-β1 serum content could account for the worse status at disease onset and thereby could help identify patients at a higher risk for disease progression. Indeed, higher TGF-β1 serum content was associated with frequent spontaneous remissions (Zissel et al. 1996). TGF-β1 has mainly an anti-inflammatory action in incipient disease, which may explain a rather protective role of this factor in the early stage of sarcoidosis. VEGF produced by monocytes at granuloma sites mirrors prolonged chronic inflammation, which may result in disease progression. FGF-2, as a mitogen for fibroblasts, may also negatively influence the disease course (Ziora et al. 2000).

In this study, in stage III of sarcoidosis a trend toward increased serum VEGF and PDGF-AB was noticed in patients with a relevantly reduced lung diffusing capacity. The ACE level was not associated with serum changes in any of the four markers tested. A subgroup of stage III patients with serum sIL-2R ≥900 U/mL presented a strong tendency toward augmented VEGF and TGF-β1 contents compared to patients with sIL-2R <900 U/mL. This finding may suggest exaggerated remodeling processes with VEGF-accentuated angiogenesis and both PDGF-AB- (Nishioka et al. 2013) and TGF-β1-induced incipient matrix accumulation in the face of prolonged inflammation. In line with these results, TGF-β1 was shown to contribute to disease progression *via* its profibrotic action.

A limitation of this study is a relatively small number of patients, especially in stage IV of sarcoidosis. This could explain, at least partly, some non-significant differences noticed between groups and subgroups. However, most of other studies have also been performed in small groups without the participation of stage IV patients (Piotrowski et al. 2015; Ziora et al. 2015). Further, this is a cross-sectional study showing the levels of serum markers only at one determined time point. Therefore, the bearing of conclusions on the involvement of the markers tested in disease progression may be limited. A prospective, longitudinal study design is required to provide more robust data in this regard.

In conclusions, VEGF, TGF-β1, FGF-2, and PDGF-AB seem to play an important role in the course of pulmonary sarcoidosis. Proangiogenic VEGF, and profibrotic FGF-2 and PDGF-AB may contribute to the progression of this disorder, whereas TGF-β1 with its dual anti-inflammatory/ profibrotic action may be a protective or deleterious factor depending on the disease stage. Reduced lung diffusing capacity and elevated serum levels of ACE and sIL-2R are potential predictors of disease worsening.

Acknowledgments We gratefully acknowledge Ms. Sabine Ring for her excellent technical assistance. We would like to thank all patients who participated in the study and the Sarkoidose-Netzwerk for its support.

Conflicts of Interest The authors declare no conflicts of interest in relation to this article.

References

Baughman RP, Winget DB, Bowen EH, Lower EE (1997) Predicting respiratory failure in sarcoidosis patients. Sarcoidosis Vasc Diffuse Lung Dis 14:154–158

Fireman E, Gilburd D, Marmor S (2009) Angiogenic cytokines in induced sputum of patients with sarcoidosis. Respirology 14:117–123

Hauber HP, Gholami D, Meyer A, Pforte A (2003) Increased interleukin-13 expression in patients with sarcoidosis. Thorax 58:519–524

Iannuzzi MC, Fontana JR (2011) Sarcoidosis: clinical presentation, immunopathogenesis, and therapeutics. JAMA 305:391–399

Jonth AC, Silveira L, Fingerlin TE, Sato H, Luby JC, Welsh KI, Rose CS, Newman LS, du Bois RM, Maier LA, ACCESS Group (2007) TGF-beta 1 variants in chronic beryllium disease and sarcoidosis. J Immunol 179:4255–4262

Judson MA (1999) An approach to the treatment of pulmonary sarcoidosis with corticosteroids: the six phases of treatment. Chest 115:1158–1165

King TE Jr, Tooze JA, Schwarz MI, Brown KR, Cherniack RM (2001) Predictingsurvival inidiopathic pulmonary fibrosis: scoring system and survival model. Am J Respir Crit Care Med 164:1171–1181

Limper AH, Colby TV, Sanders MS, Asakura S, Roche PC, DeRemee RA (1994) Immunohistochemical localization of transforming growth factor-beta 1 in the nonnecrotizing granulomas of pulmonary sarcoidosis. Am J Respir Crit Care Med 149:197–204

Marie JC, Liggitt D, Rudensky AY (2006) Cellular mechanisms of fatal early-onset autoimmunity in mice with the T cell-specific targeting of transforming growth factor-beta receptor. Immunity 25:441–454

Marshall BG, Wangoo A, Cook HT, Shaw RJ (1996) Increased inflammatory cytokines and new collagen formation in cutaneous tuberculosis and sarcoidosis. Thorax 51:1253–1261

Nardi A, Brillet PY, Letoumelin P, Girard F, Brauner M, Uzunhan Y, Naccache JM, Valeyre D, Nunes H (2011) Stage IV sarcoidosis: comparison of survival with the general population and causes of death. Eur Respir J 38:1368–1373

Newman LS, Rose CS, Maier LA (1997) Sarcoidosis. N Engl J Med 336:1224–1234

Nishioka Y, Azuma M, Kishi M, Aono Y (2013) Targeting platelet-derived growth factor as a therapeutic approach in pulmonary fibrosis. J Med Investig 60:175–183

Noor A, Knox KS (2007) Immunopathogenesiss of sarcoidiosis. Clin Dermatol 25:250–258

Patterson KC, Franek BS, Müller-Quernheim J, Sperling AI, Sweiss NJ, Niewold TB (2013) Circulating cytokines in sarcoidosis: phenotype-specific alterations for fibrotic and non-fibrotic pulmonary disease. Cytokine 61:906–911

Piotrowski WJ, Kiszałkiewicz J, Górski P, Antczak A, Górski W, Pastuszak-Lewandoska D, Migdalska-Sęk M, Domańska-Senderowska D, Nawrot E, Czarnecka KH, Kurmanowska Z, Brzeziańska-Lasota E (2015) Immunoexpression of TGF-β/Smad and VEGF-A proteins in serum and BAL fluid of sarcoidosis patients. BMC Immunol 16:58

Rothe JH, Keitzer R, Röttgen R (2010) Intestinale Sarkoidose – außergewöhnliche Manifestation einer granulomatösen Systemerkrankung. Fortschr Röntgenstr 182:74–76 (Article in German)

Sekiya M, Ohwada A, Miura K, Takahashi S, Fukuchi Y (2003) Serum vascular endothelial growth factor as a possible prognostic indicator in sarcoidosis. Lung 181:259–265

Swigris JJ, Olson AL, Huie TJ, Fernandez-Perez ER, Solomon J, Sprunger D, Brown KK (2011) Sarcoidosis-related mortality in the United States from 1988 to 2007. Am J Respir Crit Care Med 183:1524–1530

Tolnay E, Wiethege T, Kuhnen C, Wulf M, Voss B, Müller KM (1997) Expression of type IV collagenase correlates with the expression of vascular endothelial growth factor in primary non-small cell lung cancer. J Cancer Res Clin Oncol 123:652–658

Willis BC, Liebler JM, Luby-Phelps K, Nicholson AG, Crandall ED, du Bois RM, Borok Z (2005) Induction of epithelial-mesenchymal transition in alveolar epithelial cells by transforming growth factor-beta1: potential role in idiopathic pulmonary fibrosis. Am J Pathol 166:1321–1332

Yamashita M, Mouri T, Niisato M, Kowada K, Kobayashi H, Chiba R, Satoh T, Sugai T, Sawai T, Takahashi T, Yamauchi K (2013) Heterogeneous characteristics of lymphatic microvasculatures associated with pulmonary sarcoid granulomas. Ann Am Thorac Soc 10:90–97

Ziora D, Dworniczak S, Niepsuj G, Neipsuj K, Jarosz W, Sielska-Sytek E, Ciekalska K, Oklek K (2000) Proangiogenic cytokines (bFGF and VEGF) in BALF from two different lung segments examined by high resolution computed tomography (HRCT) in patients with sarcoidosis. Pneumonol Alergol Pol 68:120–130

Ziora D, Jastrzębski D, Adamek M, Czuba Z, Kozielski JJ, Grzanka A, Kasperska-Zajac A (2015) Circulating concentration of markers of angiogenic activity in patients with sarcoidosis and idiopathic pulmonary fibrosis. BMC Pulm Med 15:113

Zissel G, Homolka J, Schlaak J, Schlaak M, Müller-Quernheim J (1996) Anti-inflammatory cytokine release by alveolar macrophages in pulmonary sarcoidosis. Am J Respir Crit Care Med 154:713–719

Advs Exp. Medicine, Biology - Neuroscience and Respiration (2018) 41: 67–76
https://doi.org/10.1007/5584_2018_202

Published online: 4 May 2018

Congruence Between Pulmonary Function and Computed Tomography Imaging Assessment of Cystic Fibrosis Severity

Anna Rybacka, Joanna Goździk-Spychalska, Adam Rybacki, Tomasz Piorunek, Halina Batura-Gabryel, and Katarzyna Karmelita-Katulska

Abstract

In cystic fibrosis, pulmonary function tests (PFTs) and computed tomography are used to assess lung function and structure, respectively. Although both techniques of assessment are congruent there are lingering doubts about which PFTs variables show the best congruence with computed tomography scoring. In this study we addressed the issue by reinvestigating the association between PFTs variables and the score of changes seen in computed tomography scans in patients with cystic fibrosis with and without pulmonary exacerbation. This retrospective study comprised 40 patients in whom PFTs and computed tomography were performed no longer than 3 weeks apart. Images (inspiratory: 0.625 mm slice thickness, 0.625 mm interval; expiratory: 1.250 mm slice thickness, 10 mm interval) were evaluated with the Bhalla scoring system. The most frequent structural abnormality found in scans were bronchiectases and peribronchial thickening. The strongest relationship was found between the Bhalla sore and forced expiratory volume in 1 s (FEV1). The Bhalla sore also was related to forced vital capacity (FVC), FEV1/FVC ratio, residual volume (RV), and RV/total lung capacity (TLC) ratio. We conclude that lung structural data obtained from the computed tomography examination are highly congruent to lung function data. Thus, computed tomography imaging may supersede functional assessment in cases of poor compliance with spirometry procedures in the lederly or children. Computed tomography also seems more sensitive than PFTs in the assessment of cystic fibrosis progression. Moreover, in early phases of cystic fibrosis, computed tomography, due to its excellent resolution, may be irreplaceable in monitoring pulmonary damage.

A. Rybacka (✉)
Department of Diagnostic Imaging, University Hospital of Lord's Transfiguration, Poznan, Poland
e-mail: akna1988@gmail.com

J. Goździk-Spychalska, A. Rybacki, T. Piorunek, and H. Batura-Gabryel
Department of Pulmunology, Allergology and Respiratory Oncology, Poznan University of Medical Sciences, Poznan, Poland

K. Karmelita-Katulska
Department of General Radiology, University of Medical Sciences, Poznan, Poland

Keywords
Bhalla scoring system · Bronchiectases · Computed tomography · Cystic fibrosis · Lung imaging · Pulmonary function tests

1 Introduction

Dysfunction of the pulmonary system is a major cause of morbidity and mortality in cystic fibrosis (Davis et al. 1996). Monitoring of the lung damage in cystic fibrosis is crucial in standard clinical practice and in testing new treatment strategies. The disease progression should be under strict control from the very beginning to provide patients with the most adequate therapy. Therefore, a wide range of diagnostic procedures such as computed tomography, magnetic resonance imaging, sputum culture and pulmonary function tests (PFTs) are taken into consideration in the evaluation of pulmonary status in cystic fibrosis (Tepper et al. 2016; Wielpütz et al. 2016; Ong and Ramsey 2015; Pittman 2015; Walkup and Woods 2015; Duncan and Aurora 2014; Davies and Alton 2009). The most accurate tool in the evaluation of the lung structure in cystic fibrosis is computed tomography, as demonstrates high sensitivity in detecting bronchiectases, peribronchial thickening, and mucus plugging emphysema and air trapping, all of which are the cardinal features of cystic fibrosis (Kuo et al. 2016; Murphy et al. 2016; Bonavita and Naidich 2012; Abbott et al. 2009). Computed tomography scan images also serve to tailor treatment to patient needs. A degree of diagnostic linkage between PFTs and computed tomography has been investigated, but the correspondence between the two methods is not full clear. Bhalla et al. (1991) failed to demonstrate a significant relationship between the computed tomography image score and the percentage of predicted values of forced expiratory volume in 1 s (FEV1) and forced vital capacity (FVC), while such a relationship has been demonstrated in other studies (Demirkazık et al. 2001; Helbich et al. 1999). In contradistinction, relationship between the FEV1/FVC ratio and computed tomography image score has not been demonstrated by some investigators (Demirkazık et al. 2001), while it has been by others (Bhalla et al. 1991). Therefore, the objective of this study was to re-evaluate the congruence of PFTs and computed tomography imaging, based on the Bhalla scoring, in the assessment of cystic fibrosis severity.

2 Methods

2.1 Study Population

The study was approved by the institutional Review Board for Research which waived the requirement to obtain patient consent due to the retrospective nature of the study. The patients who were enrolled into the study had confirmed clinical, radiological, and genotypic features of cystic fibrosis. Files of the patients who had computed tomography of the lungs and PFTs performed within a maximum three-week interval during December 2012 to July 2015 were singled out from the database of the Department of Diagnostic Imaging of Poznan University of Medical Sciences, Poland. The study population amounted to 40 patients (26 females and 14 males; age range 18–40 years). Body mass index (BMI), spirometry, body plethysmography, and sputum cultures were investigated at the time of computed tomography examinations.

According to the clinical recommendations in cystic fibrosis (Ramsey and Boat 1994), patients had computed tomography and PFTs on a regular basis every 2 years as part of standard follow-up. In additional, extra evaluation was performed when a patient suffered from recurring exacerbations or presented some other complications such as hemoptysis or chest pain. There were 22 patients evaluated during exacerbations in the study group.

2.2 Computed Tomography

Patients were examined in the supine position. The inspiratory scans were obtained at suspended inspiration by using spiral acquisitions with thin-slice reconstructions consisting of 0.625 mm slice thickness every 0.625 mm. The expiratory images were obtained during expiration with 1.25 mm slice thickness and 10 mm interval. The expiratory phase scans were acquired as the incremental high-resolution computed tomography (HRCT) in groups of 4 slices each, preceded by the command 'let the air out'. The examination parameters were 120 kV with the amperage adjusted individually to the patient's anatomy by the Asir system. The acquisition time was about 4 s for a standard patient. The lung window settings were: level − 600 HU and width 1500 HU. Computed tomography scans were performed with GE LightSpeed VCT64-slice computed tomography scanner (GE Healthcare Milwaukee, WI).

Computed tomography scans were retrospectively and independently scored in random order with the Bhalla scoring system and assessed by a single observer who was blinded to the patients' personal data (Bhalla et al. 1991). Each of the seven categories (severity of bronchiectases, severity of peribronchial thickening, bronchiectases extension, mucus plugging extension, sacculations or abscesses extension, generations of bronchi affected by bronchiectases or mucus plugging, and number of bullae) was scored from 0 to 3 and the last two categories (emphysema and collapse/consolidation) were scored from 0 to 2. Then the summated score was subtracted from 25 giving the final Bhalla score result. The lower the total Bhalla score the more severe was lung disease. The results were classified as excellent when the Bhalla total score result was between 21 and 25 points, good between 20 and 16 points, mild between 11 and 15 points, moderate between 6 and 10 points, and serious between 0 and 5 points.

HRCT abnormalities were defined according to the recommendations issued by the Nomenclature Committee of the Fleischner Society (Austin et al. 1996). Bronchiectases was identified when the findings included a bronchus with an internal diameter larger than its accompanying pulmonary artery, lack of tapering of the bronchial lumen in the range of more than 2 cm, and visualization of a bronchus within 1 cm of the costal pleura. Peribronchial thickening was classified as a bronchial wall thickness measuring more than 1 mm. (Fig. 1) Mucus plugging was identified by visualization of mucus plugs in large airways or the presence of tree-in-bud or centrilobular nodules in the peripheral areas of the lungs (Figs. 2 and 3). Sacculations were characterized as dilated bronchi with a cystic or saccular appearance (Fig. 4). A bulla was defined as a round focal airspace with a thin wall and a diameter larger than 1 cm (Fig. 5). Emphysema was classified by areas of decreased attenuation, with disruption of the underlying vascular pattern and absence of well-defined walls. Consolidation was identified by increased lung opacification, which obscured the underlying parenchyma.

2.3 Pulmonary Function Tests (PFTs)

Age, weight, height, and BMI were used for clinical evaluation. Pulmonary function was assessed in this study at the shortest time interval from the computed tomography scans. Over one half of patients (22 out of the 40 patients) had PFTs performed within 2 days from the date of HRCT scans. Seventeen patients had PFTs performed during a 2-week interval from the HRCT scans, and for one patient the interval between HRCT and PFTs was 20 days. PFTs results included forced expiratory volume in 1 s (FEV1), forced vital capacity (FVC), residual volume (RV), and total lung capacity (TLC), which were expressed as the percentage of predicted values according to European Community for Steel and Coal (ECSC/ERS) (Quanjer et al. 1993). The ratio of FEV1 to FVC and RV to TLC was calculated and expressed as a percentage. PFTs were carried out with a diagnostic Jaeger MasterScreen system (Erich Jaeger GmbH; Würzburg, Germany).

2.4 Statistical Analysis

Data were expressed as means ±SD. Normality of data distribution was checked with the Shapiro-Wilk test. Since the distribution was in most cases skewed, non-parametric Mann

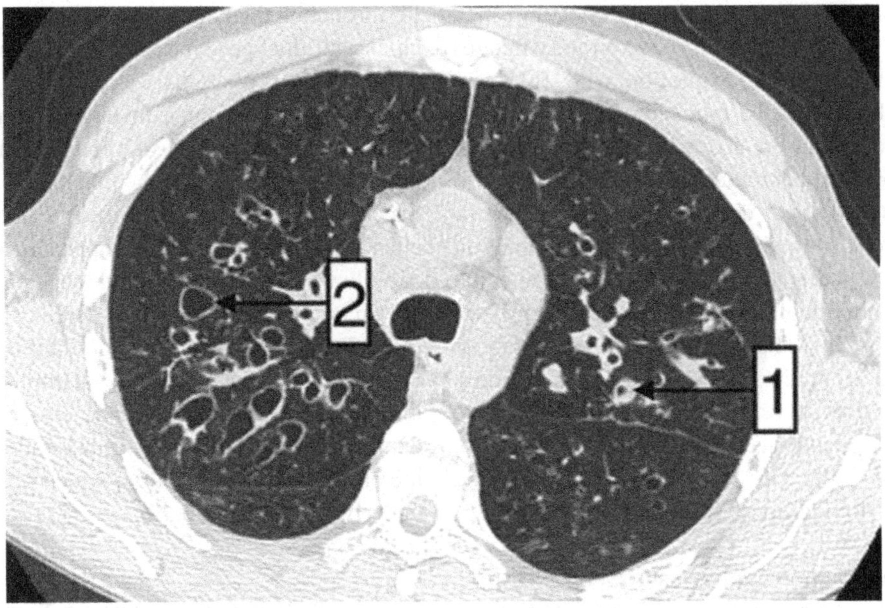

Fig. 1 A scan of a 32-year-old patient presents multiple abnormalities: bilateral peribronchial thickenings at the level of the carina (*1*) and bronchiectases with saccular appearance in the right upper lobe (*2*). The patient had normal results of pulmonary function tests

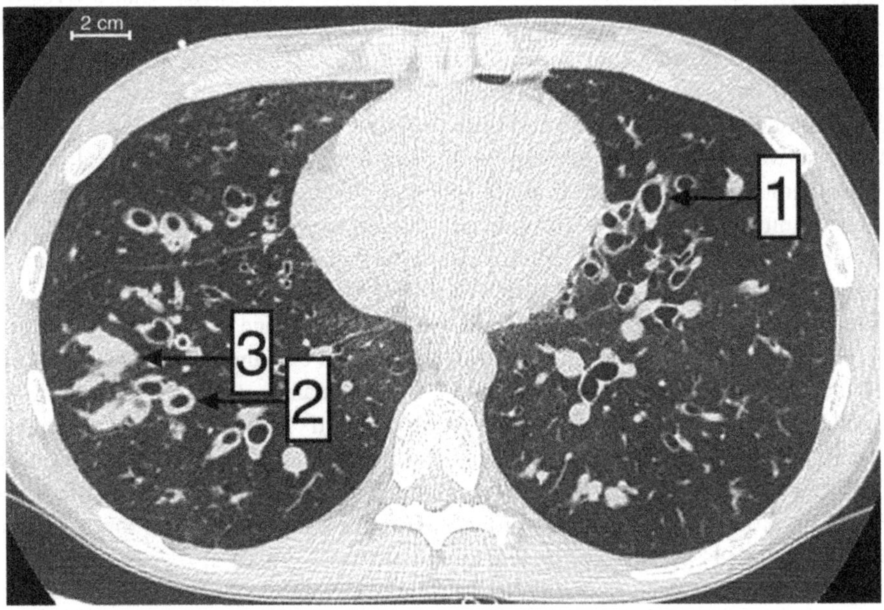

Fig. 2 Axial image of widespread bronchiectases: internal diameter of the bronchus larger than that of the accompanying pulmonary artery (*1*); peribronchial thickening – bronchial wall thickness > 1 mm (*2*); and peripheral mucus plugging (*3*)

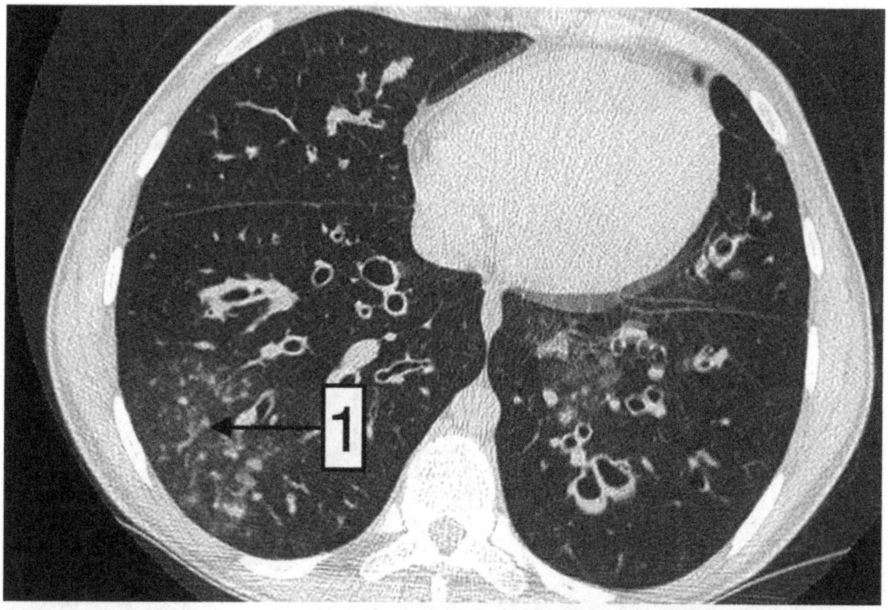

Fig. 3 Tree-in-bud sign and centrilobular nodules (*1*) caused by mucus plugging in peripheral areas of lungs

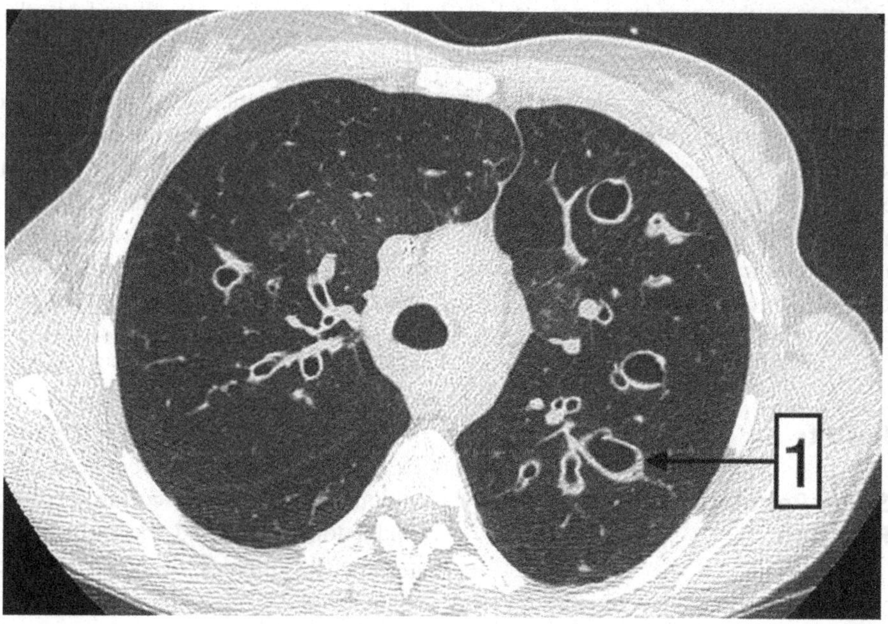

Fig. 4 Dilated bronchi with saccular appearance in the left upper lobe (*1*)

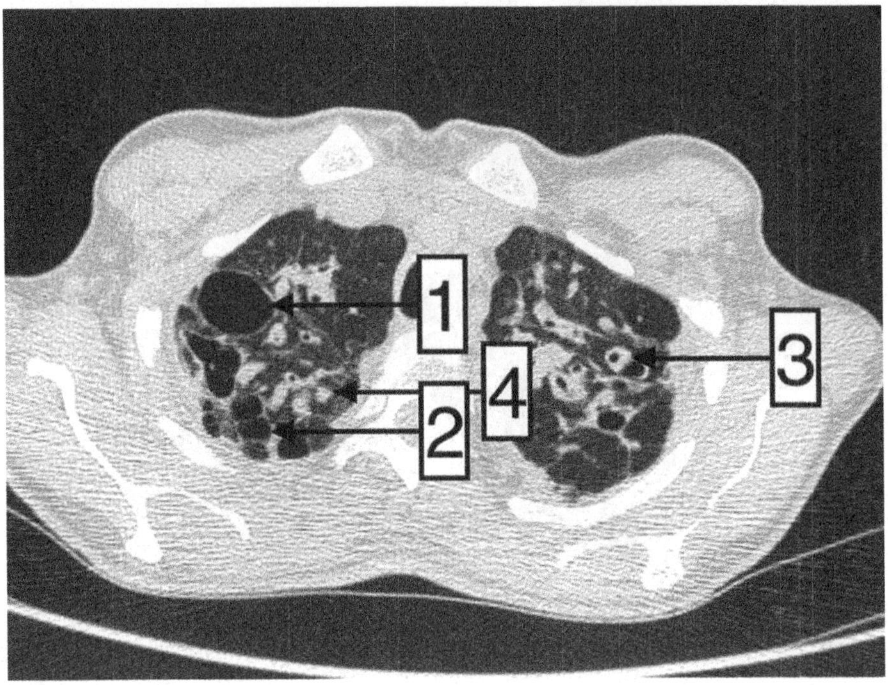

Fig. 5 Axial image of the upper lobes presents bulla in the right upper lobe (*1*), bilateral bronchiectases (*2*), peribronchial thickening (*3*), and mucus plugging (*4*)

Whitney U test was used in statistical comparisons. Correlations between computed tomography score, PFTs, and other clinical data were assessed with Spearman's rank correlation coefficient. A p-value < 0.05 defined statistically significant differences. Statistical analysis was performed using a commercial Statistica v10 package (StatSoft Inc. Tulsa, OK).

3 Results

3.1 Study Population and Pulmonary Function

Table 1 shows patient characteristics, PFTs results, and the Bhalla total score results. Twenty two patients out of the 40 enrolled into the study were evaluated during exacerbations of their condition.

Patients were divided into four groups based on the results of FEV1: <40%, 40–59%, 60–80%, and > 80%, considered as severe, moderate, mild, and normal. This division demonstrates that pulmonary function was severely defunct in over one third of patients with cystic fibrosis (Table 2). Functional impairment was in line with the severe-to-moderate intensity of cystic fibrosis expression in 58% of patients, assessed from the Bhalla score system (Table 3). The mean total Bhalla score for all patients was 10.0 ± 6.2 points; it was 7.4 ± 4.0 in patients with exacerbation and 13.2 ± 7.0 without exacerbation; the difference between these two groups of patients was significant ($p = 0.006$). There was a strong relationship between thin-section computed tomography scoring, expressed by the Bhalla total score and the pulmonary function spirometry variables (Table 4). There was no relationship between the Bhalla score and age or BMI.

3.2 Computed Tomography

Bronchiectases were the most frequent abnormality in HRCT images ($n = 38$; 95%), followed by peribronchial thickening ($n = 36$; 90%), mucus

Table 1 Baseline characteristics of the study population

	n	Mean ± SD	Min – Max
Age (year)	40	26.5 ± 6.6	18.0–40.0
BMI (kg/m^2)	40	20.4 ± 2.3	16.7–27.0
FEV1/FVC (%)	40	66.1 ± 13.4	40.6–93.9
FEV1 (L)	40	1.9 ± 1.1	0.4–4.7
FEV1 (% predicted)	40	52.3 ± 26.6	13.4–101.2
FVC (L)	40	2.7 ± 1.3	0.8–6.1
FVC (% predicted)	40	64.7 ± 24.5	24.7–109.8
TLC (L)	39	6.0 ± 1.3	3.6–9.5
TLC (% predicted)	39	108.3 ± 17.9	74.9–162.1
RV (L)	39	3.3 ± 1.4	1.3–7.6
RV% (% predicted)	39	219.1 ± 86.5	87.4–461.6
RV/TLC	39	53.9 ± 17.5	25.2–79.7
Bhalla total score (points)	40	10.0 ± 6.2	1.0–25.0

BMI body mass index, *FEV1* forced expiratory volume in one second, *FVC* forced vital capacity, *TLC* total lung capacity, *RV* residual volume

Table 2 Stratification od cystic fibrosis patients according to the level of forced expired volume in one second (FEV1)

FEV1 (%)	Patients; n (%)
<40	14 (35)
40–59	10 (25)
60–80	9 (23)
>80	7 (18)

Table 3 Bhalla total score assessing the intensity of cystic fibrosis

Bhalla score (points)	Patients; n (%)
Serious 0–5	9 (23)
Moderate 6–10	14 (35)
Mild 11–15	11 (28)
Good 16–20	2 (5)
Excellent 21–25	4 (10)

plugging (n = 34; 85%), collapse/consolidation (n = 25; 63%), and emphysema (n = 25; 63%). The abnormalities noticed less frequently were the following: sacculations and abscesses, identified in 20 (50%) and bullae in 9 (23%) patients. Over 30% of patients had a Bhalla total score between 6 and 10 points, which indicates a moderate intensity of the disease (Table 3). The exemplary pulmonary lesions are displayed in the HRCT scans presented below.

4 Discussion

Congruence of computed tomography scores and PFTs has already been under investigation in order to justify the usefulness of computed tomography scoring systems in clinical practice. Although a majority of previous studies has shown that CT scores agrees well with PFTs and with other indicators of clinical status (Pereira et al. 2014; De Jong et al. 2004b; Maffessanti et al. 1996), there is a degree of inconsistency regarding which of the PFTs is best congruent to computed tomography severity score in patients with cystic fibrosis (Demirkazık et al. 2001; Helbich et al. 1999; Bhalla et al. 1991). In this study, a highly significant congruence was observed between the Bhalla scoring system in computed tomography and FEV1, FEV1%, FVC, FVC%, FEV1/FVC ratio, RV%, and RV/TLC ratio. Notably, the strongest congruence was present for the Bhalla score and the FEV1 value which is a key PFT for the timing of lung transplantation and the single best predictor of mortality (Kerem et al. 1992).

Our present findings reflect the literature opinion that the Bhalla score is congruent to pulmonary function tests and that computed tomography scoring can be useful in monitoring patients with cystic fibrosis. PFTs are the most commonly used clinical tool to control the course of cystic fibrosis. As far as computed tomography

Table 4 Relationship between pulmonary function variables and Bhalla total score (BTS)

	n	*r*	*p*
FEV1 (L) – BTS	40	0.72	<0.001
FEV1 (% predicted) – BTS	40	0.69	<0.001
FVC (L) – BTS	40	0.69	<0.001
FVC (% predicted) – BTS	40	0.67	<0.001
FEV1/FVC – BTS	40	0.53	<0.001
TLC (L) – BTS	39	0.12	0.517
TLC (% predicted) – BTS	39	−0.13	0.446
RV (L) – BTS	39	−0.49	0.002
RV (% predicted) – BTS	39	−0.52	0.001
RV/TLC – BTS	39	−0.57	<0.001

r Spearman's rank correlation coefficent, *FEV1* forced expiratory volume in one second, *FVC* forced vital capacity, *TLC* total lung capacity, *RV* residual volume

results are congruent to PFTs, it may substitute for functional assesement in some patients, for instance, in children or elderly, who often can hardly cooperate with spirometry testing.

On the other hand, some studies underscore that structural abnormalities may be noticed in computed tomography scans in patients having normal functional results (Brody et al. 2004; Marchant et al. 2001). In this study, we also noticed the inconsistency between computed the tomography score and PFTs in individual subjects. In the seven patients having FEV1 > 80%, there were various combinations of lung lesions observable in imaging scans. Bronchiectasis was present in five, peribronchial thickening in four, mucus plugs in three, sacculations or abscesses in one, emphysema in one, and collapse/consolidation in two patients. The most striking observation was a discordance in imaging scans and PFTs results in one patient whose FEV1 was over 100% with the Bhalla score of seven, indicating severe lung damage (Fig. 1). These observations confirm previous conclusions about the incompatibility observed between function and structure. Marchant et al. (2001) have investigated clinical findings, pulmonary function, and computed tomography scans in 16 patients aged less than 12 years and noticed that all of the children had bronchiectases, including five who had normal pulmonary function measurements. Similar observations have been made by Brody et al. (2004) who found bronchiectases in imaging scans of 30% of patients with FEV1, FVC, and FEF 25–75 > 85% predicted. Those authors have presumed that computed tomography is more sensitive than PFTs in detection of lung damage in patients with cystic fibrosis. That notion has been confirmed by De Jong et al. (2004a) who assessed HRCT scans and PFTs of 48 patients, and repeated the assessment after a follow-up of 2 years. The authors employed and compared five scoring systems to evaluate the imaging scans: the Castile, Brody, Helbich, Santamaria, and Bhalla scoring. They report that all imaging scans significantly worsen while PFTs remain stable or improved over time. Similar conclusions have been reached in further studies on the subject (de Jong et al. 2006; Judge et al. 2006). The present findings are, to an extent, in line with those above mentioned showing that computed tomography may be more sensitive for the assessment of progression of cystic fibrosis than functional lung evaluation in some cases. Thus, scoring of HRCT examination is essential in managing patients with cystic fibrosis.

Apart from using a complex scoring systems, there is an alternative way to assess structural lung damage in computed tomography images, which is a quantitative post-processing. In this method, airway wall and lumen dimensions are measured using an image analysis system. De Jong et al. (2005) have assessed computed tomography scans of 23 children with cystic fibrosis at baseline and after a 2-year follow-up and compared them with the control group. These authors demonstrate a progressive airway wall thickening and bronchiectases in patients with cystic fibrosis. The usefulness of the post-processing of computed tomography images has also been validated in other investigations, showing distinct differences in lung structure in various stages of cystic fibrosis severity and a good association with PFTs variables (Santos et al. 2016; Walkup and Woods 2015; Montaudon et al. 2007).

This study has several limitations. Firstly, it is a retrospective study as imaging data were collected between 2012 and 2015. Secondly, computed tomography examinations were scored by one radiologist. In the majority of previous studies, two or more radiologists used to asses scans (Pereira et al. 2014; Brody et al. 2004; Demirkazık et al. 2001). However, recent studies show a good intra- and inter-observer agreement for the Bhalla scoring system (De Jong et al. 2004b). Thus, this issue may not have had an impacted on the present results. Finally, patients with clinical symptoms of respiratory exacerbation were not excluded from the study population. We chose to include these patents on the premise that the severity score in computed tomography associates with spirometry results in asymtomatic as well as symptomatic patients and that scoring changes associate with improvements in PFTs after treatment for pulmonary exacerbation (Robinson et al. 2001; Shah et al. 1997).

In conclusion, lung function data are highly congruent to lung structural data obtained from the computed tomography examination. However, in early phases of cystic fibrosis, due to its excellent resolution, computed tomography may be irreplaceable in monitoring pulmonary damage. Lung impairment in cystic fibrosis should be based on clinical symptoms, functional, and structural assessments, which enables the targeting and monitoring of therapy.

Conflicts of interest The authors declare no conflicts of interest in relation to this article.

References

Abbott GF, Rosado-de-Christenson ML, Rossi SE, Suster S (2009) Imaging of small airways disease. J Thorac Imaging 24:285–298

Austin JH, Muller NL, Friedman PJ, Hansell DM, Naidich DP, Remy-Jardin M, Webb WR, Zerhouni EA (1996) Glossary of terms for CT of the lungs: recommendations of the nomenclature Committee of the Fleischner Society. Radiology 200:327–331

Bhalla M, Turcios N, Aponte V, Jenkins M, Leitman BS, McCauley DI, Naidich DP (1991) Cystic fibrosis: scoring system with thin section computed tomography. Radiology 179:783–788

Bonavita J, Naidich DP (2012) Imaging of bronchiectases. Clin Chest Med 33:233–248

Brody AS, Klein JS, Molina PL, Quan J, Bean JA, Wilmott RW (2004) High-resolution computed tomography in young patients with cystic fibrosis: distribution of abnormalities and correlation with pulmonary function tests. J Pediatr 145:32–38

Davies JC, Alton EW (2009) Monitoring respiratory disease severity in cystic fibrosis. Respir Care 54:606–617

Davis PB, Drumm M, Konstan MW (1996) Cystic fibrosis. Am J Respir Crit Care Med 154:1229–1256

de Jong PA, Nakano Y, Lequin MH, Mayo JR, Woods R, Paré PD, Tiddens HA (2004a) Progressive damage on high resolution computed tomography despite stable lung function in cystic fibrosis. Eur Respir J 23:93–97

de Jong PA, Ottink MD, Robben SG, Lequin MH, Hop WC, Hendriks JJ, Paré PD, Tiddens HA (2004b) Pulmonary disease assessment in cystic fibrosis: comparison of CT scoring systems and value of bronchial and arterial dimension measurements. Radiology 231:434–439

de Jong PA, Nakano Y, Hop WC, Long FR, Coxson HO, Paré PD, Tiddens HA (2005) Changes in airway dimensions on computed tomography scans of children with cystic fibrosis. Am J Respir Crit Care Med 172:218–224

de Jong PA, Lindblad A, Rubin L, Hop WC, de Jongste JC, Brink M, Tiddens HA (2006) Progression of lung disease on computed tomography and pulmonary function tests in children and adults with cystic fibrosis. Thorax 61:80–85

Demirkazık FB, Arıyürek OM, Özçelik U, Gocmen A, Hassanabad HK, Kiper N (2001) High resolution CT in children with cystic fibrosis: correlation with pulmonary functions and radiographic scores. Eur J Radiol 37:54–59

Duncan JA, Aurora P (2014) Monitoring early lung disease in cystic fibrosis: where are we now? Breathe 10:34–47

Helbich TH, Heinz-Peer G, Eichler I, Wunderbaldinger P, Götz M, Wojnarowski C, Brasch RC, Herold CJ (1999) Cystic fibrosis: CT assessment of lung involvement in children and adults. Radiology 213:537–544

Judge EP, Dodd JD, Masterson JB, Gallagher CG (2006) Pulmonary abnormalities on high-resolution CT demonstrate more rapid decline than FEV1 in adults with cystic fibrosis. Chest 130:1424–1432

Kerem E, Reisman J, Corey M, Canny GJ, Levison H (1992) Prediction of mortality in patients with cystic fibrosis. N Engl J Med 326:1187–1191

Kuo W, Kemner-van de Corput MP, Perez-Rovira A, de Bruijne M, Fajac I, Tiddens HA (2016) Multicentre chest computed tomography standardisation in children and adolescents with cystic fibrosis: the way forward. Eur Respir J 47:1706–1717

Maffessanti M, Candusso M, Brizzi F, Piovesana F (1996) Cystic fibrosis in children: HRCT findings and distribution of disease. J Thorac Imaging 11:27–38

Marchant JM, Masel JP, Dickinson FL, Masters IB, Chang AB (2001) Application of chest high-resolution computer tomography in young children with cystic fibrosis. Pediatr Pulmonol 3:24–29

Montaudon M, Berger P, Cangini-Sacher A, de Dietrich G, Tunon- de-Lara JM, Marthan R, Laurent F (2007) Bronchial measurement with three-dimensional quantitative thin-section CT in patients with cystic fibrosis. Radiology 242:573–581

Murphy KP, Maher MM, O'Connor OJ (2016) Imaging of cystic fibrosis and pediatric bronchiectasis. AJR Am J Roentgenol 206:448–454

Ong T, Ramsey BW (2015) Update in cystic fibrosis 2014. Am J Respir Crit Care Med 192:669–675

Pereira FF, Ibiapina Cda C, Alvim CG, Camargos PA, Figueiredo R, Pedrosa JF (2014) Correlation between Bhalla score and spirometry in children and adolescents with cystic fibrosis. Rev Assoc Med Bras 60:216–221

Pittman JE (2015) Assessment and detection of early lung disease in cystic fibrosis. Pediatr Allergy Immunol Pulmonol 28:212–219

Quanjer PH, Tammeling GJ, Cotes JE, Pedersen OF, Peslin R, Yernault JC (1993) Lung volumes and forced ventilatory flows report working party standardization of lung function tests, European community for steel and coal official statement of the European respiratory society. Eur Respir J 16(Suppl):5–40

Ramsey BW, Boat TF (1994) Outcome measures for clinical trials in cystic fibrosis: summary of a cystic fibrosis foundation consensus conference. J Pediatr 124:177–192

Robinson TE, Leung AN, Northway WH, Blankenberg FG, Bloch DA, Oehlert JW, Al-Dabbagh H, Hubli S, Moss RB (2001) Spirometer-triggered high-resolution computed tomography and pulmonary function measurements during an acute exacerbation in patients with cystic fibrosis. J Pediatr 138:553–559

Santos MK, Cruvinel DL, de Menezes MB, Teixeira SR, Vianna EO, Jorge Elias JJ, Martinez JA (2016) Quantitative computed tomography analysis of the airways in patients with cystic fibrosis using automated software: correlation with spirometry in the evaluation of severity. Radiol Bras 49:351–357

Shah RM, Sexauer W, Ostrum BJ, Fiel SB, Friedman AC (1997) High-resolution CT in the acute exacerbation of cystic fibrosis: evaluation of acute findings, reversibility of those findings, and clinical correlation. AJR Am J Roentgenol 169:375–380

Tepper LA, Ciet P, Caudri D, Quitter AL, Utens EM, Tiddens HA (2016) Validating chest MRI to detect and monitor cystic fibrosis lung disease in a pediatric cohort. Pediatr Pulmonol 51:34–41

Walkup LL, Woods JC (2015) Advances in imaging cystic fibrosis lung disease. Pediatr Allergy Immunol Pulmonol 28:220–229

Wielpütz MO, Eichinger M, Biederer J, Wege S, Stahl M, Sommerburg O, Mall MA, Kauczor HU, Puderbach M (2016) Imaging of cystic fibrosis lung disease and clinical interpretation. Rofo 188:834–845

Advs Exp. Medicine, Biology - Neuroscience and Respiration (2018) 41: 77–82
https://doi.org/10.1007/5584_2018_230

Published online: 11 September 2018

Detection of Influenza in the Epidemic Season 2016/2017 Based on I–MOVE+ Project

E. Hallmann-Szelińska, K. Cieślak, K. Szymański, D. Kowalczyk, M. R. Korczyńska, I. Paradowska-Stankiewicz, and L. B. Brydak

Abstract

In the influenza season 2016/2017 in Europe, the predominant virus was A/H3N2/. In Poland, the percentage of people vaccinated against influenza was 3.33%. European I–MOVE+ project shows how important it is to monitor the effectiveness of influenza vaccine. The project demonstrates that a match between the circulating vaccine strains and those included in the vaccine for the Northern Hemisphere was low-to-moderate. In the present study, there were 379 patients and 296 control subjects examined in hospitals in Poland as part of I–MOVE+ project. The real-time reverse transcription polymerase chain reaction (qRT-PCR) method was used to detect the influenza virus in all subjects. We detected the influenza subtype A/H3N2/ in 59.1% and type B virus in 2.1%. There was one co–infection of subtype A/H3N2/ with subtype A/H1N1/ and eight co–infections of type B with subtype A/H3N2/. No influenza viruses were detected in the control group. Only 19 patients and 22 control subjects were vaccinated during the epidemic season in question. A proportion of people vaccinated against influenza in Poland remains dismally low compared to other European countries.

Keywords

Epidemic season · Influenza · Laboratory testing · Subtype · Vaccine

1 Introduction

In 2016/2017 season in Poland subtype A/H3N2/ was dominant, while the major etiological agent of influenza-like infection was the respiratory syncytial virus. In the season, a three-component vaccine was in use, with inactivated split and subunit antigen of the following composition: A/California/7/2009 (H1N1) pdm09 – like virus, A/Hong Kong/4801/2014 (H3N2) – like virus, and B/Brisbane/60/2008 – like virus Victoria lineage (Grohskopf et al. 2017). In this season, the percentage of vaccinated people in Poland remained at a very low level of 3.33% general population.

The I-MOVE (Influenza – Monitoring Vaccine Effectiveness) project, aimed at measuring

E. Hallmann-Szelińska (✉), K. Cieślak, K. Szymański, D. Kowalczyk, and L. B. Brydak
Department of Influenza Research, National Influenza Center, National Institute of Public Health – National Institute of Hygiene, Warsaw, Poland
e-mail: ehallmann@pzh.gov.pl

M. R. Korczyńska and I. Paradowska-Stankiewicz
Department of Epidemiology, National Institute of Public Health – National Institute of Hygiene, Warsaw, Poland

influenza vaccine effectiveness in Europe, coordinated by EpiConcept and ECDC (ECDC 2017) has been conducted in European countries since 2007. Poland has started to participate in the project since the influenza season 2010/2011 (Kissling et al. 2012). The aim of the present report is to identify and evaluate the activity of influenza virus and to determine the effectiveness of vaccination against influenza in the 2016–2017 influenza season as part of the I-MOVE project implemented in Poland.

2 Methods

This study was approved by the Review Board for Human Research at the National Influenza Center of the National Institute of Public Health – National Institute of Hygiene in Warsaw, Poland. It was conducted in accordance with the principles for human research set by the Declaration of Helsinki.

2.1 Patients and Material Sampling

To this case-control study, patients were qualified who met the EU influenza-like illness (ILI) case definition. The definition is as follows: sudden onset of symptoms, at least one of the following four systemic symptoms: fever or feverishness, malaise, headache, and myalgia, and at least one of the following three respiratory symptoms: cough, sore throat, and shortness of breath (Kissling et al. 2017). The material for the study consisted of pharyngeal and nasal swabs taken from patients whose clinical presentation met the definition of ILI. Forty six general practitioners (GP) voluntarily participated in the collection of study material in 14 voivodships of Poland. Data for laboratory qRT-PCR investigation were collected from 7 age groups: 0–4, 5–9, 10–14, 15–25, 26–44, 45–64, and 65+ years. The control group consisted of patients with ILI symptoms, but having a negative result of influenza qRT-PCR test from the pharyngeal and nasal swabs. Each sample was sent to the reference laboratory in the Department of Influenza Research – National Influenza Center of the National Institute of Public Health – National Institute of Hygiene in Warsaw with attached information regarding the date of a GP visit and swabbing, and the age and gender of a participating patient. The material sampled, submitted by GP was stored at −80 °C until laboratory virus identification.

2.2 Real-Time Reverse Transcription (qRT-PCR)

RNA Isolation Influenza virus RNA was isolated using a set of Maxwell 16 Viral Total Nucleic Acid Purification Kit (Promega Corporation, Madison, WI) with 200 μl of a clinical sample suspended in phosphate buffered saline (PBS), according to the manufacturer's instructions for low elution volume (LEV) cartridges. The RNA was suspended in 50 μl of RNase-free water.

Influenza A virus subtypes were determined by qRT-PCR with a light cycler thermocycler 2.0 (Roche Diagnostics; Rotkreuz, Switzerland). qRT-PCR analysis was performed in capillary tubes with a capacity of 20 μl using 0.5 μl (20 nM) primers and 0.5 μl (5 nM) probes for each of the reaction. Primers and probes were developed in the Influenza Reagent Resource (IRR) program run by the US Center for Disease Control (CDC). The reaction mixture containing buffer, $MgSO_4$, bovine serum albumin (BSA), RNase free water, and SuperScript® III/platinum *Taq* mix (Invitrogen by Life Technologies – Thermo Fisher Scientific, Carlsband, CA) was incubated with 5 μl of RNA in each capillary tube. As a positive control, reference strains of influenza virus in the vaccine for 2016/2017 epidemic season were used, and negative control consisted of RNase-free water. Before DNA amplification, template RNA was reverse transcribed to obtain the corresponding cDNA. Reverse transcription was performed at 50 °C for 30 min. The DNA was then subjected to the initial denaturation (1 cycle at 95 °C for 2 min), with further amplification steps comprising:

denaturation at 95 °C for 15 s, annealing at 55 °C for 30 s, and elongation at 72 °C for 20 s; repeated in 45 cycles (Hallmann-Szelińska et al. 2016).

2.3 Vaccine Effectiveness

Influenza vaccine effectiveness (IVE) is defined as the percentage of risk reduction in the vaccinated individuals relative to the unvaccinated individuals. Vaccine effectiveness was estimated according to the following formula: IVE = (1-OR) × 100. The OR is calculated by dividing the vaccination rate in the case group by the vaccination rate in the control group. It refers to the risk of influenza in the vaccinated group compared to the unvaccinated group (Weinberg and Szilagyi 2010). In addition, other factors such as age and co-morbidities can influence vaccine effectiveness.

3 Results and Discussion

In the epidemic season 2016/2017, a group of patients with positive results of influenza qRT-PCR test consisted of 379 patients: 208 females and 171 males. There were 296 subjects: 178 females and 118 males, in the control group with negative results of the test. Influenza type A was detected in 362 (95.5%) samples, in which subtype A/H3N2/ predominated in 214 (59.1%), while the remaining cases of type A were unsubtyped. One co–infection of A/H1N1/ with A/H3N2/ and eight of A/H3N2/ with influenza type B were reported. The infections caused by type B virus amounted to 2.1% (Fig. 1). Data were collected from patients in seven age groups, thanks to the voluntary participation of 46 general practitioners across 15 Polish regions. Only were 19 patients with positive PCR test results in the study group and 22 in the control group vaccinated during the reported period (Table 1).

Virological characteristics of samples within I-MOVE+ project, collected by various GPs and in various hospitals during the epidemic season 2016/2017 in Poland, distributed by the week number confirmed that the A type virus was the main circulating culprit. The collection of samples started in 45th ISO week of 2016. The pick of this season was in 3rd ISO week of the

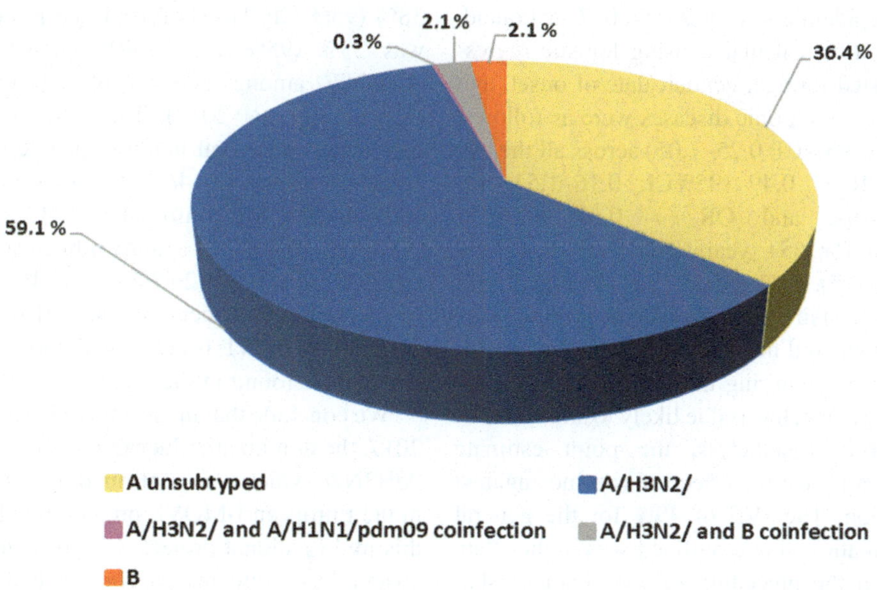

Fig. 1 Proportion of influenza viruses detected in the patient group investigated within the I-MOVE project

Table 1 Descriptive display of influenza A and B patients and control subjects evaluated in the framework of I-MOVE+ project during the epidemic season 2016–2017 in Poland (379 patients and 296 controls)

		Patients	Controls
		n (%)	n (%)
Age groups (year):	0–4	29 (7.7)	29 (9.8)
	5–9	25 (6.6)	9 (3.0)
	10–14	25 (6.6)	4 (1.4)
	15–25	35 (9.2)	37 (12.5)
	26–44	119 (31.4)	109 (36.8)
	45–64	94 (24.8)	84 (28.4)
	65+	52 (13.7)	24 (8.1)
Sex	Male	171 (45.1)	118 (39.9)
	Female	208 (54.9)	178 (60.1)
Vaccination status: 2016–2017 season	Vaccinated	19 (5.0)	22 (7.4)
	Unvaccinated	360 (95.0)	274 (92.6)
Vaccination status: 2014–2015 season	Vaccinated	13 (3.5)	14 (4.7)
	Unvaccinated	356 (96.5)	282 (95.3)
	Missing	10 (2.6)	0 (0)
Influenza type	A/H1N1/pdm09	0 (0)	–
	A/H3N2/	224 (59.1)	–
	A unsubtyped	138 (36.1)	–
	A/H3N2/ + A/H1N1/pdm09	1 (0.3)	–
	A/H3N2/ + B	8 (2.1)	–
	B	8 (2.1)	–
Chronic conditions, including obesity and pregnancy	Yes	103 (27.2)	67 (22.6)
	No	276 (72.8)	229 (77.4)

following year, which speaks for the typical course of the epidemic curve (Fig. 2).

In the epidemic season 2016/2017 in Poland, the odds ratios, calculated using logistic regression, adjusted for age, gender, date of onset, and the presence of chronic diseases were as follows: OR = 0.51 (95%CI: 0.25–1.05) across all the age groups, OR = 0.49 (95%CI: 0.16–1.51) for 15–64 years, and OR = 0.67 (95%CI: 0.16–1.99) for 65+ years The IVE was 49% (95%CI: 7–75), 51% (95%CI: 51–84), and 33% (95%CI: 9–84) in the respective age groups. Vaccination status did not have a statistically significant effect on acquiring the infection in any age group tested, which was due likely to small size of age groups. Nonetheless, the point estimate suggested a protective effect of vaccines against the infection. The IVE of 49% for the general adult population above outlined was higher than the 21% in the preceding season (Paradowska-Stankiewicz et al. 2018). For comparison, in other European studies conducted in much larger cohorts of people, consisting of several thousand people, IVE in the same 2016/2017 season was 38% (95% CI: 21–51) for all age groups, and it was 23% (95% CI: 15–49) against influenza A/H3N2/ among persons 65+ years of age (Kissling et al. 2017). Likewise, in a study conducted in several hundred people during the epidemic season 2014/2015, the overall pooled and adjusted for confounders IVE was 47.5% (95% CI: 16.4–67.0) against influenza A(H1N1) pdm2009, and it was 29.7% (95% CI: -34.4–63.2) for A(H3N2) (Valenciano et al. 2015). Our present estimate of IVE is in line with those moderate-to-low IVE found in those studies.

We conclude that in the epidemic season 2016/2017, the dominant influenza virus in Poland was A/H3N2/, which was confirmed by data obtained in the European I-MOVE project. It is hoped that this multi-national project will promote a greater awareness of influenza vaccination that despite all the efforts made so far remains at a dismally low levels in the lay public. The present formula used

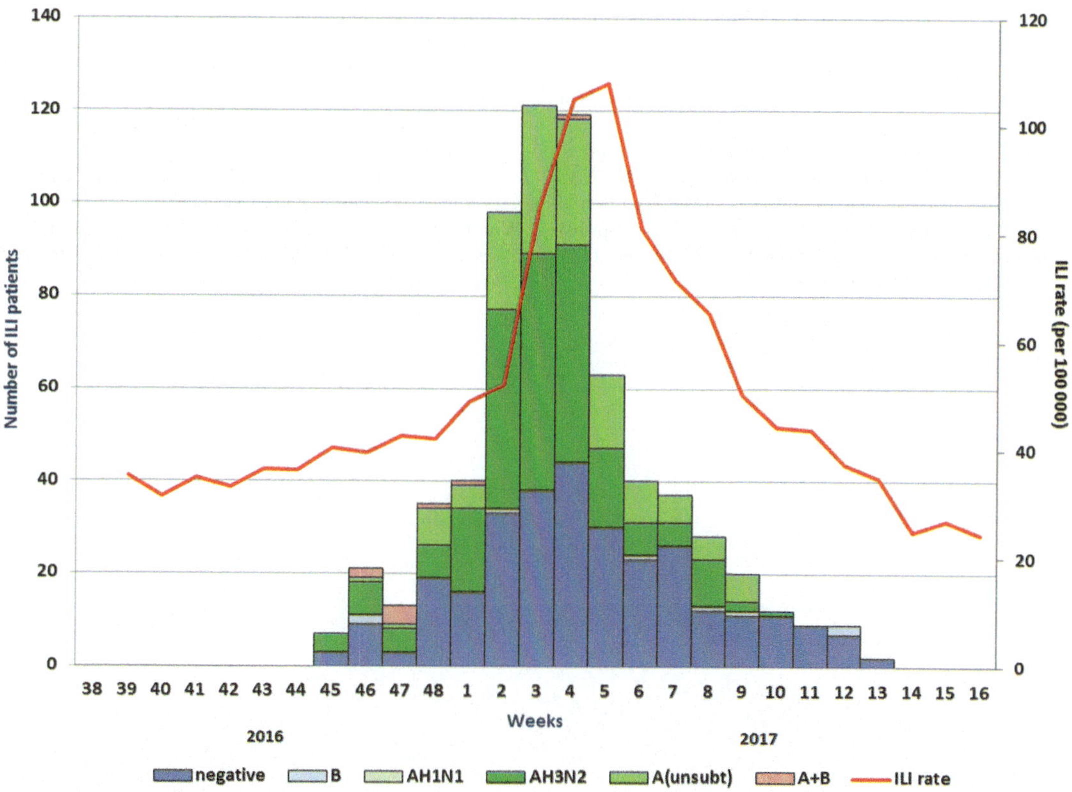

Fig. 2 Virological characteristics of samples within the I-MOVE + project distributed by the week number, in the ISO week date standard, during the epidemic season 2016–2017 in Poland. ILI, influenza-like illness

to estimate the influenza vaccine effectiveness may require modifications as it seems not to exactly mirror a drop in risk of acquiring the infection after influenza vaccination, even taking into account the studied group size and possible confounders, which may be both misleading and disenchanting in undertaking vaccination.

Acknowledgments This work was funded by grant 634446 within the project: "I-MOVE+ Integrated Monitoring of Vaccines Effects in Europe: a platform to measure and compare effectiveness and impact of influenza and pneumococcal vaccines and vaccination strategies in the elderly – I-MOVE-plus". Additionally, funded in part by the NIPH-NIH theme 3/EM. We would like to acknowledge physicians and employees of the Voivodship Sanitary Epidemiological Stations in Poland who participated in this program.

Conflicts of Interest The authors declare no conflicts of interest in relation to this article.

References

ECDC (2017) European centre for disease prevention and control. https://sites.google.com/site/epiflu/. Accessed 26 Apr 2018

Grohskopf LA, Sokolow LZ, Broder KR, Walter EB, Bresee JS, Fry AM, Jernigan DB (2017) Prevention and control of seasonal influenza with vaccines: recommendations of the advisory committee on immunization practices – United States, 2017–18 influenza season. MMWR Recomm Rep 66(2):1–20

Hallmann-Szelińska E, Bednarska K, Korczyńska MR, Paradowska-Stankiewicz I, Brydak LB (2016) Virological characteristics of the 2014/2015 influenza season based on molecular analysis of biological material derived from I-MOVE study. Adv Exp Med Biol 921:81–85

Kissling E, Valenciano M, I-MOVE Case-Control Studies Team (2012) Early estimates of sesonal influenza vaccine effectiveness in Europe among target groups for vaccination: results from the I-MOVE multicentre case-control study, 2011/12. Euro Surveill 17(15): pii:20146

Kissling E, Rondy M, I-MOVE/I-MOVE+ study team (2017) Early 2016/17 vaccine effectiveness estimates against influenza A(H3N2): I-MOVE multicentre case control studies at primary care and hospital levels in Europe. Euro Surveill 22(7):pii:30464

Paradowska-Stankiewicz I, Korczyńska MR, Cieślak K, Kowalczyk D, Szymański K, Brydak LB (2018) Vaccine effectiveness against influenza in 2015/16 in hospital and ambulatory medical care facilities: polish results of the European I-MOVE+ multicenter study. Adv Exp Med Biol 1023:93–100

Valenciano M, Kissling E, Reuss A, Jiménez-Jorge S, Horváth JK, Donnell JM, Pitigoi D, Machado A, Pozo F, I-MOVE Multicentre Case Control Study Team (2015) The European I-MOVE multicentre 2013–2014 case-control study. Homogeneous moderate influenza vaccine effectiveness against A(H1N1) pdm09 and heterogeneous results by country against A (H3N2). Vaccine 33(24):2813–2822

Weinberg GA, Szilagyi PG (2010) Vaccine epidemiology: efficacy, effectiveness, and the translational research roadmap. J Infect Dis 201:1607–1610

Advs Exp. Medicine, Biology - Neuroscience and Respiration (2018) 41: 83–87
https://doi.org/10.1007/5584_2018_229

Published online: 17 August 2018

Influenza and Influenza-Like Viruses: Frequent Infections in Children Under 14 Years of Age During the 2016/2017 Epidemic Season

K. Cieślak, D. Kowalczyk, K. Szymański, E. Hallmann-Szelińska, and L. B. Brydak

Abstract

Influenza is an infectious disease that is a common cause of infection among children. The main reason for it is the extremely low percentage of vaccinated people in Poland. In the 2016/2017 epidemic season more than 3,000 tests from children up to age 14 were examined. The dominance of subtype A/H3N2/ (40.9%) was confirmed. The evaluation was stratified by three age groups (0–4, 5–9, and 10–14 years), which revealed significant differences. The highest number of samples was available in the 0–4 years group. The highest percentage of positive samples was present in the 10–14 years group. Influenza-like viral infections, among them the respiratory syncytial virus, were also observed. Children due to immature immunity are at particular risk for influenza. A lack of proper vaccination coverage strongly increases the chance of serious complications of the infection.

K. Cieślak (✉), D. Kowalczyk, K. Szymański, E. Hallmann-Szelińska, and L. B. Brydak
Department of Influenza Research, National Influenza Centre, National Institute of Public Health- National Institute of Hygiene, Warsaw, Poland
e-mail: kcieslak@pzh.gov.pl

Keywords

Children · Epidemic season · Infection · Influenza virus · Molecular biology · Respiratory tract

1 Introduction

In each epidemic season in Poland we observe the occurrence of a seasonal influenza epidemic. Influenza is an infectious disease in which children are particularly vulnerable because of the not fully formed immune system (Hallmann-Szelińska et al. 2015). In the epidemic season 2016/2017, the dominant subtype circulating in Poland was the A/H3N2/ virus.

The best way to prevent illnesses among children is seasonal vaccination. In the epidemic season 2016/2017, according to the recommendations of the American Academy of Pediatrics (AAP) vaccination in children over 6 months of age and older was recommended, as well as in children with increased risk of complications caused by influenza. It also is important to vaccinate people who are in contact with children under 5 years of age. It is recommended to vaccinate pregnant women (2 and 3 trimesters), which makes newborn babies less vulnerable to contract the infection (Jackowska 2016). Unfortunately, the percentage of vaccinated children in Poland is still dramatically low. In 2016, 0.60% of children aged 0–4 years and 0.94% children aged 5–14 years were vaccinated (Epimeld 2017).

In case of influenza infection, especially in children, rapid diagnosis is important as it allows the introduction of antiretroviral treatment which should start up to 48 h after onset of symptoms (Brydak 2008). The most precise and efficient method of confirming influenza virus in a test sample consists of molecular biology, be it reverse transcription polymerase chain reaction (RT-PCR) or real-time quantitative transcription polymerase chain reaction (qRT-PCR) (Cieślak et al. 2017). However, the method is as yet hardly available in tertiary hospitals. Instead, rapid antigen diagnostic tests are often used (Nitsch-Osuch et al. 2017). Their use requires less time, which is an important factor in diagnosis (Chartrand et al. 2012). Rapid methods of influenza diagnosis significantly reduce the overuse of antibiotics in children (Nitsch-Osuch et al. 2016). The neuraminidase inhibitors oseltamivir and zanamivir are the only causal treatment of children infected by influenza virus. Administration of zanamivir is possible only in in-hospital setting, as the drug is given by inhalation (Jackowska 2016).

2 Methods

This study was approved by an institutional Ethics Committee and it was conducted in accordance with the Declaration of Helsinki for Human Research. The analysis concerned patients under 14 years of age. The study material consisted of samples of pharyngeal and nasal swabs samples taken from patients in the whole country within the framework of the Sentinel and Non-Sentinel surveillance systems. Specimens were collected in both Department of Influenza Research of the National Influenza Center in Warsaw as well as in laboratories of 16 Voivodship Sanitary and Epidemiological Stations across Poland.

RNA isolation from the influenza virus was performed using the Maxwell 16 Viral Total Nucleic Acid Purification Kit (Promega Corp., Madison, WI), according to the manufacturer's instructions of. A 200 μl clinical sample, suspended in phosphate buffered saline, was used. The RNA was suspended in 50 μl of RNase-free water.

Molecular analysis was performed by real-time qRT-PCR using the Light Thermocycler 2.0 System (Roche Diagnostics; Rotkreuz, Switzerland). Influenza A and B subtypes were identified using primers and probes obtained from the International Reagent Resource (IRR) program run by the US Center for Disease Control (CDC). The analysis was carried out in capillary tubes with a capacity of 20 μl, with 0.5 μl (20 nM) primers and 0.5 μl (5 nM) probes for each of the reaction. Briefly, reverse transcription was performed at 50 °C for 30 min. The DNA was then subjected to the initial denaturation (1 cycle at 95 °C for 2 min), with further amplification steps comprising the following: denaturation at 95 °C for 15 s, annealing at 55 °C for 30 s, and elongation at 72 °C for 20 s; repeated in 45 cycles (for details see Hallmann-Szelińska et al. 2016). Positive control consisted of the influenza virus strains: A/California/7/2009 (H1N1)pdm09, A/Hong Kong/4801/2014(H3N2), and B/Brisbane/60/2008, included in the vaccine elaborated for the epidemic season 2016/2017. The negative control consisted of RNase- free water.

For the confirmation of the genetic material of respiratory viruses such as influenza A and B viruses, respiratory syncytial virus (RSV) A and B, human metapneumovirus (HMPV), human adenovirus, human coronaviruses (HCoV)-229E/NL63 and OC43/HKU1, parainfluenza viruses (PIV): PIV-1, PIV-2, PIV-3, and PIV-4, enteroviruses (EV), human bocavirus 1/2/3/4. RT-PCR reaction was performed using RV15 OneStep ACE Detection Kit (Seeplex; Seoul, South Korea), according to the manufacturer's instructions. The product was detected by agarose gel electrophoresis.

3 Results and Discussion

In the 2016/2017 epidemic season, 4,094 samples were analyzed in the whole country, including 1,096 tests from children under 14 years of age. The highest number of 755 samples was in the 0–4 age group, of which 24.6% tested positive. In the children of 5–9 years, positive and negative samples accounted for 50.2% and 49.8%, respectively. In contrast, in the 10–14 age group, the proportion was reversed, compared with the youngest age group, the majority of 61.0% samples tested positive (Fig. 1).

The predominant influenza virus found was the A/H3N2/ subtype. This subtype accounted for 40.9% of all confirmed cases of influenza in children under 14 years of age. Influenza A unsubtyped infections were confirmed in 39.8% of samples, while influenza B only in 2.3%. Influenza-like viruses were also observed, accounting for 17.1% (Fig. 2). The most common was respiratory syncytial virus (73 cases). Occasional cases of parainfluenza 2 and 3 viruses, and rhinoviruses were detected (Table 1).

When stratified by the age group, the highest number of cases of influenza A virus (unsubtyped) and subtype A/H3N2/ was found in the children aged 0–4 years; there were 111 and 74 infections caused by either subtype, respectively (Fig. 3). In the remaining two age groups, an appreciably number of confirmed cases of influenza was observed. There were 38 cases caused by the influenza A unsubtype virus and 58 cases caused by the A/H3N2/ subtype in the children aged 5–9 years and 28 cases of influenza A unsubtype and 50 cases of the A/H3N2/ subtype in those aged 10–14. Influenza type B was rather rare in the 2016/2017 season; there were overall just 10 cases reported. The most of them (7) were observed among children aged 5–9. No cases of influenza A/H1N1/ and influenza A/H1N1/pdm09 were reported.

The prevalence of the influenza A/H3N2/ subtype in children up to the age of 14 years in Poland during the 2016/2017 epidemic season was confirmed. These data coincide with those provided by other European countries (WHO 2017). This is in contradistinction to the preceding 2015/2016 season when the predominant variant was rather the A/H1N1/pdm09 subtype and when, in the latter part of the season, frequent influenza B types were observed; the finding unseen in Poland in 2016/2017 (Cieślak et al. 2017).

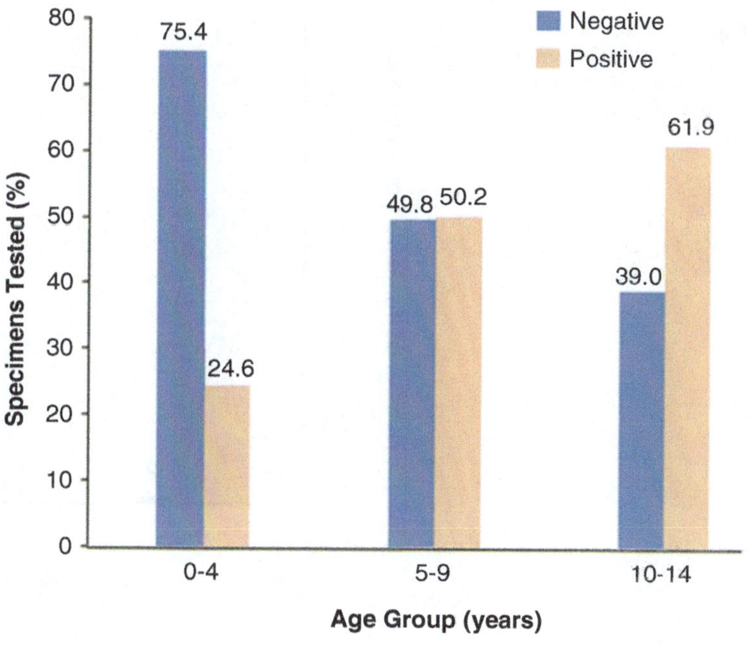

Fig. 1 Percentage of positive and negative specimens tested distributed by age group in the 2016/2017epidemic season

Fig. 2 Percentage of influenza and influenza-like infections stratified by age groups in the 2016/2017 epidemic season

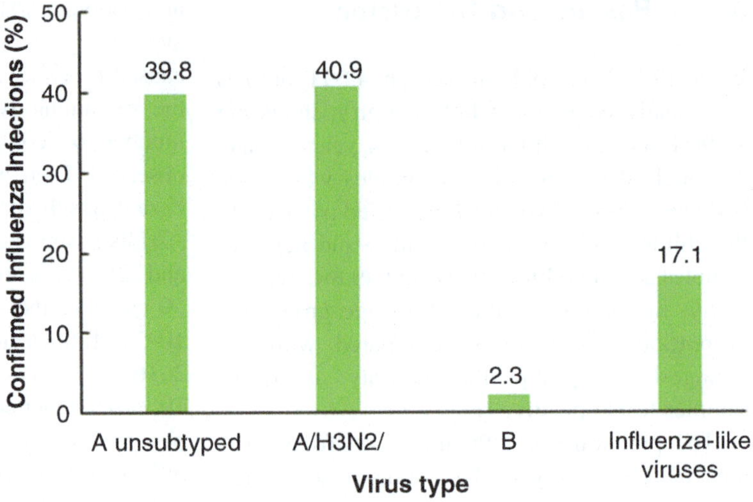

Table 1 Confirmations of influenza-like viruses in different age-groups in the 2016/2017 epidemic season

Age group (years)	ADV	RSV	PIV1	PIV2	PIV3	HCoV	HMPV	Total
0–4	0	69	0	0	1	0	0	70
5–9	0	3	0	0	0	0	0	3
10–14	0	1	0	1	0	0	0	2
Total	0	73	0	1	1	0	0	75

ADV Aleutian disease virus, *RSV* respiratory syncytial virus, *PIV* parainfluenza viruses: PIV-1, PIV-2, PIV-3, *HCoV* human coronaviruses, *HMPV* human metapneumovirus

Fig. 3 Number of confirmed cases influenza stratified by age group in the epidemic season 2016/2017

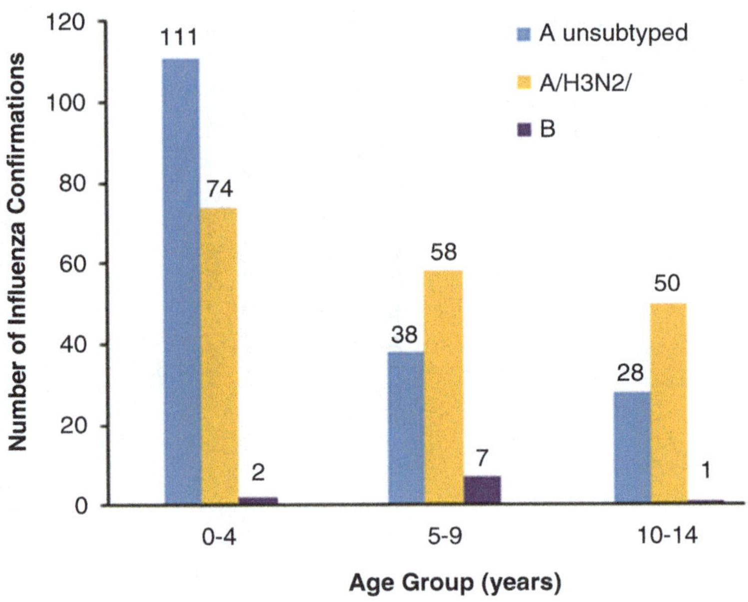

In children aged 0–14 years, 445 cases of influenza and influenza- like viruses were confirmed. Such a high number of cases reflects the widespread problem of low vaccination rate in Poland. Since the child's immune system is not full matured, complications can often easily occur, which may lead serious morbidity or even or death (Życińska and Brydak 2007).

We revealed an interesting trend in all the age groups investigated. The older the children's age, the fewer there were confirmed cases of influenza. That was due liable to fewer pediatric consultations and, consequently, missed cases of infection. In the youngest children of 0–4 years the confirmed cases of influenza were distinctly more common (755) than in the oldest children of 10–14 years (136).

There was a lower incidence of respiratory syncytial virus compared to the preceding 2015/2016 season, but the predomination of this virus among all influenza–like viruses was consistent (Cieślak et al. 2017). This confirms the notion that respiratory syncytial virus is the most common cause of lower respiratory tract infection among children. In 2005, 33.8 million cases of confirmed infection with respiratory syncytial virus have been reported in children under 5 years of age worldwide (Gerretsen and Sande 2017).

Acknowledgments Funded by the NIP- NIH thematic subject 3/EM. The authors acknowledge a help from physicians and employees of the Voivodship Sanitary and Epidemiological Stations engaged in the SENTINEL program of influenza surveillance in Poland.

Conflicts of Interest The authors declare no conflicts of interest in relation to this article.

References

Brydak LB (2008) Influenza, pandemic flu, myth or real threat? Rythm, Warsaw 1-492 (in Polish)

Chartrand C, Leeflang MM, Minion J, Brewer T, Pai M (2012) Accuracy of rapid influenza diagnostic tests: a meta-analysis. Ann Intern Med 156:500–511

Cieślak K, Szymański K, Kowalczyk D, Brydak LB (2017) Influenza and influenza-like viruses in children aged under 14 years in the epidemic season 2015/2016 in Poland. Adv Exp Med Biol 968:13–18

Epimeld (2017) Preventive vaccinations in Poland. Available from: http://wwwold.pzh.gov.pl/oldpage/epimeld/index_p.html#05. Accessed 10 Apr 2018

Gerretsen HE, Sande CJ (2017) Development of respiratory syncytial virus (RSV) vaccines for infants. J Infect 74(Suppl 1):S143–S146

Hallmann- Szelińska E, Bednarska K, Brydak LB (2015) Influenza in children aged under 14 years in Poland. Probl Hig Epidemiol 96(3):535–539

Hallmann-Szelińska E, Bednarska K, Korczyńska MR, Paradowska-Stankiewicz I, Brydak LB (2016) Virological characteristics of the 2014/ 2015 influenza season based on molecular analysis of biological material derived from I-MOVE study. Adv Exp Med Biol 921:81–85

Jackowska T (2016) Recommendations for prevention of influenza in children, 2016–2017. Pediatr Pol 91 (6):606–613

Nitsch-Osuch A, Gyrczuk E, Wardyn A, Życinska K, Brydak LB (2016) Antibiotic prescription practices among children with influenza. Adv Exp Med Biol 905:25–31

Nitsch-Osuch A, Kuchar E, Gołębiak I, Kanecki K, Tarka P, Brydak LB (2017) Rapid influenza diagnostic tests improve suitability of antiviral treatment in hospitalized children. Adv Exp Med Biol 968:1–6

WHO (2017) http://apps.who.int/flumart/Default?ReportNo=1. Accessed 25 July 2017

Życińska K, Brydak LB (2007) Influenza and flu prophylaxis – a current medical issue. Pol Arch Intern Med 117(10):464–469

Advs Exp. Medicine, Biology - Neuroscience and Respiration (2018) 41: 89
https://doi.org/10.1007/978-3-030-03870-0

Index

The manufacturer's authorised representative in the EU is Springer
Nature Customer Service Centre GmbH, Europaplatz 3, 69115 Heidelberg,
Germany. If you have any concerns regarding our products, please
contact ProductSafety@springernature.com

Printed and bound by CPI Group (UK) Ltd, Croydon, CR0 4YY
30/09/2026
02225259-0019